Ernst Schering Foundation Symposium
Proceedings 2006-3
New Avenues to Efficient Chemical Synthesis

Ernst Schering Foundation Symposium
Proceedings 2006-3

New Avenues to Efficient Chemical Synthesis

Emerging Technologies

P.H. Seeberger,
T. Blume
Editors

With 144 Figures

Springer

centers around the world as well as in many industrial laboratories in pharmaceutical and fine chemical industries.

2) Nanotechnology and Catalysis Research in Synthesis
Nanotechnology is an evolving research area especially in materials and biotechnological sciences. First studies have shown that the special properties of nanoparticles can give rise to highly active and selective catalysts to enable chemists to perform entirely novel transformations. Discussion and evaluation of the potential of nanoparticles for chemical research in a pharmaceutical company with experts in the field was needed. Other areas in catalysis like biotransformations and metal catalyst screening and development continue to expand the possibilities for the manufacturing of test compounds and development candidates.

3) Rapid Microwave Assisted Organic Synthesis
Another synthesis technology which has just started to impact and change the way chemical synthesis is performed in many laboratories is microwave assisted organic synthesis. Using microwave reactors, reaction times often can be reduced from hours or days to minutes or even seconds. Selectivities and yields often can be increased drastically. Therefore, this technology has the potential to increase the output of chemical drug discovery units enormously. An important question in this field is how to scale up these transformations in microwave reactors up to kilogram scale.

4) New Developments in Solid Supported Synthesis
Solid-supported technologies are already well established methods in medicinal chemistry and automated synthesis. Over the last couple of years new trends have evolved in this field which are of utmost importance as they have the potential to revolutionize the way chemical synthesis especially for library production is performed. Microchip-based synthesis technologies and multistep sequences with solid-supported catalysts or reagents in flow-through systems are only two spectacular examples. A new approach is the use of solid-supported systems for the scale-up of chemical reactions thereby enabling the rapid and smooth transition from discovery to development units.

The importance of chemical synthesis as a core technology in the pharmaceutical and chemical industry, prompted us to organize an Ernst Schering Foundation Workshop in 2006 to bring together scientific leaders from both academia and industry to discuss the implication of new technologies for drug discovery and development. The lectures presented at the symposium are collected in this book. The editors hope that this volume will prove useful to chemists in the pharmaceutical industry as well in research institutions and universities interested in chemical synthesis for drug discovery and development.

P.H. Seeberger
T. Blume

Contents

List of Editors and Contributors

Editors

Seeberger, P.H.
Laboratory for Organic Chemistry,
Swiss Federal Institute of Technology (ETH) Zürich, HCI F 315,
Wolfgang-Pauli-Str. 10, 8093 Zürich, Switzerland
(e-mail: seeberger@org.chem.ethz.ch)

Blume, T.
Process Research B / Automated Process Optimization,
Global Chemical Development, Bayer Schering Pharma AG,
Müllerstr. 178, 13342 Berlin, Germany
(e-mail: thorsten.blume@schering.de)

Contributors

Allmendinger, Th.
Chemical and Analytical Development, Novartis Pharma AG,
4002 Basel, Switzerland

Baxendale, I.R.
Innovative Technology Center (ACS), Department of Chemistry,
University of Cambridge, Cambridge, CB21EW, United Kingdom
(e-mail: irb21@cam.ac.uk)

Grigg, R.
Chemistry Department, Leeds University, Leeds LS2 9JT,
United Kingdom

Beller, M.
Leibniz-Institut für Katalyse e.V. University of Rostock,
Albert-Einstein-Straße 29a, 18059 Rostock, Germany
(e-mail: matthias.beller@catalysis.de)

Codée, J.D.C.
Laboratory for Organic Chemistry,
Swiss Federal Institute of Technology (ETH) Zürich, HCI F 315,
Wolfgang-Pauli-Str. 10, 8093 Zürich, Switzerland

Geyer, K.
Laboratory for Organic Chemistry,
Swiss Federal Institute of Technology (ETH) Zürich, HCI F 315,
Wolfgang-Pauli-Str. 10, 8093 Zürich, Switzerland

Haswell, St. J.
Department of Chemistry, The University of Hull,
Hull, HU6 7RX, United Kingdom
(e-mail: S.J.Haswell@Hull.ac.uk)

Jensen, K.
Department of Chemical Engineering,
Masschusetts Institute of Technology,
77 Massachusetts Avenue, Cambridge, MA 02139, USA
(e-mail: kfjensen@mit.edu)

Lehmann, H.
Novartis Instiute for Biomedical Research,
Discovery Technologies/Preparation Laboratories,
Klybeckstrasse 191, 4002 Basel, Switzerland
(e-mail: hansjoerg.lehmann@novartis.com)

Ley, S.V.
Innovative Technology Center (ACS), Department of Chemistry,
University of Cambridge, Cambridge, CB21EW, United Kingdom
(e-mail: svl1000@cam.ac.uk)

Mak, Ch.-P.
Chemical and Analytical Development, Novartis Pharma AG,
4002 Basel, Switzerland

Matteo, J.C.
217 FenceRail Gap, Walland, Tennessee, TN 37886, USA

Meisenbach, M.
Chemical and Analytical Development, Novartis Pharma AG,
4002 Basel, Switzerland
(e-mail: mark.meisenbach@novartis.com)

Sridharan, V.
Chemistry Department, Leeds University, Leeds LS2 9JT,
United Kingdom

Tao, J.
BioVerdant, Inc., 7330 Carroll Road, San Diego, CA 92121, USA
(e-mail: alex.tao@BioVerdant.com)

Warrington, B.H.
45, The Drive, Hertford, SG14 3DE, United Kingdom
(e-mail: brianwarrington@btconnect.com)

White, J.D.
William James House, Cowley Road, Cambridge CB4 0WX,
United Kingdom

Wong-Hawkes, S.Y.F.
4 Hull Close, Cheshunt, EN7 6XG, United Kingdom

Zhang, X.
Department of Chemistry, The University of Hull,
Hull, HU6 7RX, United Kingdom

Zhao, L.
BioVerdant, Inc., 7330 Carroll Road, San Diego, CA 92121, USA

Ernst Schering Foundation Symposium Proceedings, Vol. 3, pp. 1–19
DOI 10.1007/2789_2007_025

Published Online: 23 May 2007

Microreactors as Tools in the Hands of Synthetic Chemists

P.H. Seeberger(✉), K. Geyer, J.D.C. Codée

Laboratory for Organic Chemistry, Swiss Federal Institute of Technology (ETH) Zurich, HCI F 315, Wolfgang-Pauli-Str. 10, 8093 Zurich, Switzerland
email: *seeberger@org.chem.ethz.ch*

Abstract. Recent developments in the construction of microstructured reaction devices and their wide-ranging applications in many different areas of chemistry suggest that microreactors may significantly impact the way chemists conduct experiments. Miniaturizing reactions offers many advantages for the synthetic organic chemist: high-throughput scanning of reaction conditions, precise control of reaction variables, the use of small quantities of reagents, increased safety parameters, and ready scale-up of synthetic procedures. A wide range of single and multiphase reactions has been performed in microfluidic-based devices. Certainly, microreactors cannot be applied to all chemistries yet and microfluidic systems also have disadvantages. Limited reaction time ranges, high sensitivity to precipitating products, and analytical challenges have to be overcome. An overview of microfluidic devices available for chemical synthesis is provided and some specific examples, mainly from our laboratory, are discussed to illustrate the potential of microreactors.

1 Introduction

Synthetic chemists typically perform transformations in round-bottomed flasks on a scale ranging from several milligrams to many grams, in reaction volumes from less than one milliliter to several liters (Geyer et al. 2006; Flögel et al. 2006). The optimization of chemical transformations consumes considerable amounts of starting materials and often takes a lot in order to identify ideal reaction conditions. Having found the optimal conditions to achieve a certain reaction on a small scale, process scale-up often poses additional challenges and requires further adjustment of the reaction parameters. To overcome these hurdles in synthetic chemistry, microstructured continuous-flow reactors and chip-based microreactors are becoming increasingly popular. The term "microreactor" is generally used to describe microstructured reactors. In fact, the reactors are often a lot bigger than the term "microreactor" suggests, having internal volumes of several milliliters. So far microstructured reactors have found most applications in process and development chemistry. Chip-based microreactors incorporate much smaller channels and are therefore more suitable for synthetic purposes in an academic setting. The small dimensions of microreactors allow for the use of minimal amounts of reagent under precisely controlled conditions and make it possible to rapidly screen reaction conditions and improve the overall safety of the process. To obtain synthetically useful amounts of material, the reactors are simply run longer, the so-called scale-out principle (Thayer 2005). Alternatively, several reactors are placed in parallel (numbering up), ensuring identical conditions for the analytical and preparative modes. Reviews on different aspects of microreactor chemistry have appeared (Thayer 2005; Ehrfeld et al. 2000; Fletcher et al. 2002; DeWitt 1999; Jähnisch et al. 2004; Watts and Haswell 2005; Pennermann et al. 2004; Jensen 2001; Kiwi-Minsker and

Renken 2005; Hessel et al. 2005a, b; Golb and Hessel 2004; Kockmann et al. 2006). Here, some concepts and recent developments relevant to synthetic chemists will be summarized. Microreactor technology is beginning to be used in the fine chemical and pharmaceutical industries (Hessel et al. 2005b), and several success stories have been reported already (Thayer 2005; Rouhi 2004).

2 Microfluidic Reactor Types

Microreactors consist of a network of miniaturized channels, sometimes embedded in a flat surface, referred to as the chip (Ehrfeld et al. 2000; Cooper et al. 2001; Fletcher and Haswell 1999). A variety of reactors have been explored and some reactors are now commercially available. Different applications call for different types of reactors. The size as well as the chemical and physical properties of the material used for reactor construction and the mode of reagent and solvent introduction to the system influence the field of use for different reactors. The features of some microreactors are summarized in Fig. 1. [A wide variety of other microreactors is available from other manufacturers and suppliers, e.g., Institute for Microtechnology Mainz (IMM), Fraunhofer Alliance for Modular Microreaction Systems (FAMOS), the New Jersey Center for MicroChemical Systems (NJCMCS), the MicroChemical Process Technology Research Association (MCPT). The selection presented here is by no means exhaustive and only serves to indicate the diversity in systems developed to date.] Glass, silicon, stainless steel, metals, and polymers have been used to construct microreactors (Ehrfeld et al. 2000; Manz et al. 1991). Glass has traditionally been the most popular material for synthetic chemists since it is chemically inert to most reagents and solvents. The transparency of glass enables the visual inspection of reactions and fabrication procedures are well established (Watts and Haswell 2005; McCreedy 2000; Fletcher et al. 1999). The commercially available AFRICA microreactor system for example can be equipped with glass reactors of different sizes (60 μl, 250 μl, and 1.0 ml) (Syrris 2006).

Silicon has also found widespread use in the construction of microreactors, since methods developed for semiconductor chip production can

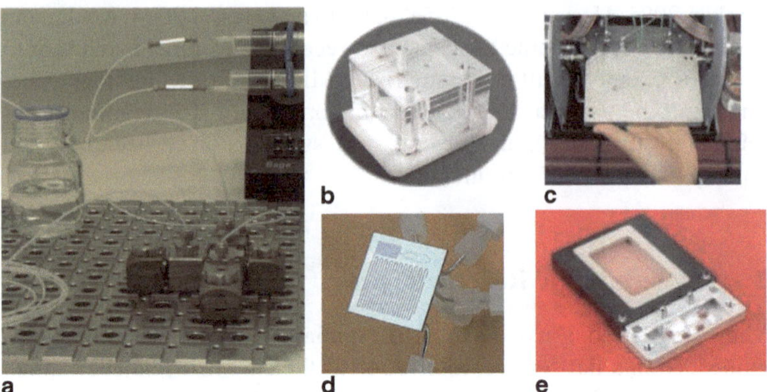

Fig. 1a–e. Selected microreactors. **a** Stainless steel microreactor system designed by Ehrfeld Mikrotechnik. **b** Glass microreactor (Watts and Haswell 2005). **c** Stainless steel microreactor of the CYTOS Lab system (http://www.cpc-net.com/cytosls.shtml). **d** Silicon-based microreactor designed by Jensen (Ratner et al. 2005). **e** Glass microreactor of the AFRICA System

be readily applied to create a variety of three-dimensional architectures (Jensen 2001; Lu et al. 2001; de Mas et al. 2002). Oxidized silicon, like glass, is chemically inert to most reagents and solvents. The excellent thermal conductivity of silicon allows for microreactors to be constructed with outstanding heat-transfer capabilities. Exothermal reactions as well as reactions that require very high or low temperatures benefit from silicon microreactors (Ratner et al. 2005).

Stainless steel is the material of choice for process chemistry. Consequently, stainless steel microreactors have been developed that include complete reactor process plants and modular systems. Reactor configurations have been tailored from a set of micromixers, heat exchangers, and tube reactors. The dimensions of these reactor systems are generally larger than those of glass and silicon reactors. These meso-scale reactors are primarily of interest for pilot-plant and fine-chemical applications, but are rather large for synthetic laboratories interested in reaction screening. The commercially available CYTOS Lab system (CPC 2007), offers reactor sizes with an internal volume of 1.1 ml and 0.1 ml, and modular microreactor systems (internal reactor volumes 0.5 ml to

11.0 ml) are offered by Ehrfeld Mikrotechnik (Ehrfeld Mikrotechnik BTS 2007). The different types of microreactors for this reactor system include capillary and cartridge reactors. IMM provides a variety of stainless steel microreaction systems including micromixers, heat exchangers and reactors for multiphase reactions.

Polymer-based microreactor systems [e.g., made of poly(dimethylsiloxane) (PDMS)], with inner volumes in the nanoliter to microliter range (Hansen et al. 2006), are relatively inexpensive and easy to produce. Many solvents used for organic transformations are not compatible with the polymers that show limited mechanical stability and low thermal conductivity. Thus the application of these reactors is mostly restricted to aqueous chemistry at atmospheric pressure and temperatures for biochemical applications (Hansen et al. 2006; Wang et al. 2006; Duan et al. 2006).

3 Flow Types and Introduction of Solvents and Reagents

Fluids can be moved through the channels of microreactors using hydrodynamic pumping, electrokinetic pumping, or capillary flow. The most straightforward approach to operate a microreactor from the synthetic chemist's point of view is to drive solutions through the reactor by hydrodynamic pumping. Flow speeds from microliters per minute to liters per minute can be achieved by using either syringe or HPLC pumps. In glass microreactors, fluids can be immobilized using electroosmotic flow (EOF), whereby a voltage is applied to the reagent and collection reservoirs. This mode of "pumping" has certain advantages over hydrodynamic pumping as it involves no moving parts and can be readily miniaturized and carefully computer controlled. However, EOF requires polar solvents, depends on the solute concentration, may cause unwanted electrochemical transformations and/or separations and can only be applied to analytical scale reactors. Capillary flow techniques are also of limited use in organic synthesis. Although precise control over fluid amounts can be achieved, the volumes transported through the capillary reactors are very small (DeWitt 1999; Nikbin and Watts 2004).

Independent of the mode of pumping, the flow inside the channel network of chip-based microreactors is generally laminar and the mixing of reagents occurs by diffusion. Given the small dimensions of the devices, diffusion is generally very efficient, and mixing is effected within milliseconds.

4 Synthesis in Microchemical Systems

Due to the small dimensions and the increased surface-to-volume ratio of microreactors, mass and heat transport are significantly more efficient than in the classical round-bottomed flask. The mixing of reagents by diffusion is very fast, and heat exchange between the reaction medium and reaction vessel is highly efficient. As a result, the reaction conditions in a continuous-flow microchannel are homogenous and can be precisely controlled. Highly exothermic and even explosive reactions can be readily harnessed in a microreactor. The careful control of reaction temperature and residence time has a beneficial effect on the outcome of a reaction with respect to yield, purity, and selectivity. Below, some examples, mainly from our laboratory, are described to illustrate the potential application of microreactors to organic synthesis. The application of microreactors to small-scale medicinal and academic total synthesis is just beginning.

5 Liquid Phase Reactions

A wide range of liquid phase reactions have been performed in microreactor devices, such as Grignard reactions (Taghavi-Moghadam et al. 2001), nitrations (Rouhi 2004; Doku et al. 2001), glycosylations (Ratner et al. 2005; Geyer and Seeberger 2007), olefinations (Skelton et al. 2001a,b; Snyder et al. 2005), peptide-couplings (Watts et al. 2001, 2002a,b) aldol reactions (Wiles et al. 2002), epoxidations (Snyder et al. 2005), multicomponent reactions (Acke et al. 2006), and Swern-oxidations (Kawaguchi et al. 2005), Liquid phase reactions carried out in microstructured devices benefit from the efficient mass and heat transport characteristics of microreactors and the fact that only small amounts of the reactants are in the system at any given time. In general, reac-

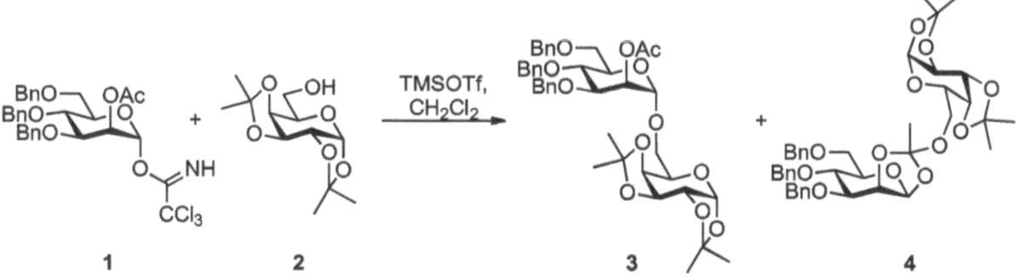

Scheme 1. Glycosylations in a microreactor

tions performed in microreactors should be reasonably fast and should not produce precipitating products or by-products that might clog up the reaction channel.

Our laboratory focused initially on the use of microreactors for reaction optimization of glycosylation reactions (Scheme 1). In general, glycosylations are very challenging, since their stereochemical outcome depends on a wide variety of factors, such as the nature of the coupling partners, temperature, solvent, and concentration. Furthermore, the building blocks used for oligosaccharide assembly often require multistep syntheses and are precious synthetic intermediates themselves. Using a Jensen silicon microreactor system, the reaction progress of the coupling between mannoside **1** and galactoside **2** was monitored as a function of temperature and time (Ratner et al. 2005). It was revealed that at low temperatures (–80°C to –70°C) and short reaction times (<1 min) the formation of orthoester **4** was favored, whereas higher temperatures (–40°C) and longer reaction times (~4 min) led to the formation of the desired α-linked product **3**. Using as little as 100 mg of starting materials, 40 different reaction conditions were scanned within 1 day. The crude reaction mixtures were analyzed off-line using LCMS.

6 Liquid–Solid Phase Reactions

Chemical transformations requiring solid starting materials, intermediates, or products are difficult to carry out in microreactors since solids may clog the channel network and hamper the continuous flow. In order to carry out reactions that use solid catalysts, several different ap-

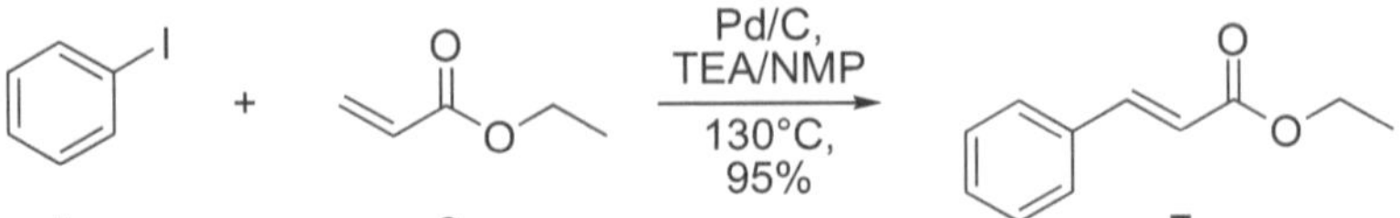

Scheme 2. Heck reaction forming ethyl cinnamate **7** in a stainless steel flow-through reactor

proaches have been reported. Catalytically active metals may be used to cover the inner walls of a reactor or may be placed on miniaturized poles in the reactor channels (Jensen 2001; Greenway et al. 2000). Alternatively, catalysts can be loaded on polymer beads in prepacked reaction cartridges that are placed in the reactor channel (Snyder et al. 2005; Saaby et al. 2005; Haswell et al. 2001b). Independent of the means of immobilization, all approaches benefit from a very high surface-to-volume ratio, characteristic of the microreactor platform. Highly effective interaction of the phases leads to considerable reaction rate enhancements.

Metal-catalyzed cross-couplings are key transformations for carbon–carbon bond formation. The applicability of continuous-flow systems to this important reaction type has been shown by a Heck reaction carried out in a stainless steel microreactor system (Snyder et al. 2005). A solution of phenyliodide **5** and ethyl acrylate **6** was passed through a solid-phase cartridge reactor loaded with 10% palladium on charcoal (Scheme 2). The process was conducted with a residence time of 30 min at 130°C, giving the desired ethyl cinnamate **7** in 95% isolated yield. The batch process resulted in 100% conversion after 30 min at 140°C using a preconditioned catalyst.

7 Liquid–Gas Reactions

Synthetic transformations using corrosive and toxic gases are generally difficult to perform because of the hazardous and highly reactive nature of the gases. Specially designed liquid–gas microreactors allow for the careful control of gas flow in the reactor and to regulate the contact time between gas and liquid. Integrated gas–liquid separators can be introduced to separate the gaseous phase at the end of the reaction (Jähnisch et al. 2004). The utility of microreactors for this chemistry has been well illustrated for fluorinations (de Mas et al. 2002, 2003; Chambers et al. 1999, 2001, 2005; Hessel et al. 2000; Jähnisch et al. 2000) chlorinations (Ehrich et al. 2002; Wehle et al. 2000), nitrations (Antes et al. 2001), and oxygenations.

The direct fluorination of toluene was performed at room temperature in a silicon microreactor that was internally coated with nickel to render it compatible with the corrosive gas (de Mas et al. 2002). Fluorinations using elemental fluorine are highly exothermic and difficult to control using conventional equipment. Taking full advantage of the highly efficient mass and heat transport as well as the presence of only a small amount of fluorine at any given time in the reactor, mono-fluorination of toluene was achieved with very good selectivity.

8 Liquid–Gas–Solid Reactions

Multiphase catalytic reactions, such as catalytic hydrogenations and oxidations are important in academic research laboratories and chemical and pharmaceutical industries alike. The reaction times are often long because of poor mixing and interactions between the different phases. The use of gaseous reagents itself may cause various additional problems (see above). As mentioned previously, continuous-flow microreactors ensure higher reaction rates due to an increased surface-to-volume ratio and allow for the careful control of temperature and residence time.

Reductive aminations are key transformations en route to many drug substances. However, reversibility, functional group incompatibility, and over-reduction can create problems. The reduction of aryl imines often gives rise to secondary amines that are contaminated with the corresponding primary amine, due to over-reduction. A commercially available hydrogenation reactor (H-cube), which combines continuous-flow microchemistry with on-demand hydrogen generation, allows for the catalytic reduction of imines with high chemoselectivity (Scheme 3) (Saaby et al. 2005; Thalesnano 2007).

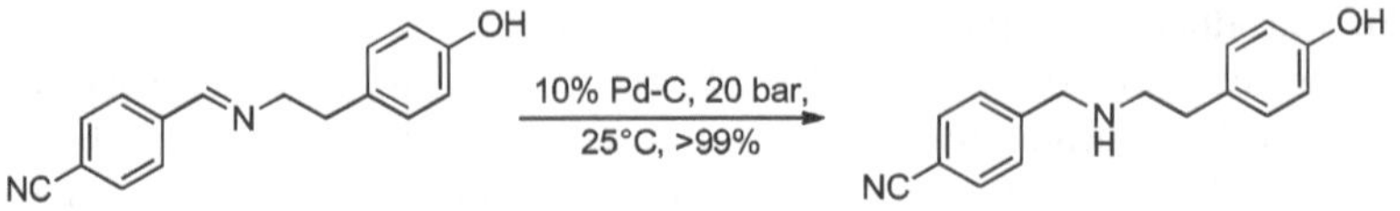

Scheme 3. Chemoselective hydrogenation of imine **8** in a hydrogenation reactor

9 Synthesis of Imaging Agents

Multistep syntheses using interconnected microreactors will eventually be a way to create complex molecules in a flow-through mode. It also allows the use of unstable intermediates that can be generated in the first reactor and then immediately fed into the second one. Some multistep syntheses using microreactor technology have already been carried out. The radiolabeled imaging probe (2-deoxy-2-[^{18}F]fluoro-*D*-glucose) ([^{18}F]FDG) **12** was prepared in a poly(dimethylsiloxane) (PDMS)-based microreactor, whereby all reaction steps were conducted in one single device (Scheme 4) (Lee et al. 2005). A highly sophisticated, tailor-made chip was designed that sequentially executed the following steps: (1) concentration of the dilute [^{18}F]fluoride solution using anion exchange column techniques; (2) solvent change, water to acetonitrile; (3) nucleophilic substitution of the mannosyl triflate **10**; (4) solvent change, acetonitrile to water; (5) acidic hydrolysis of the acetate protecting groups to obtain the [^{18}F]FDG **12**. The entire process was automated and required 14 min, where the existing batch process takes 50 min. This acceleration is significant when taking the half-life of [^{18}F]fluorine, 110 min, into account. [^{18}F]FDG **12** was produced in 38% radiochemical yield and radiochemical purity of 97.6%, and was directly used for

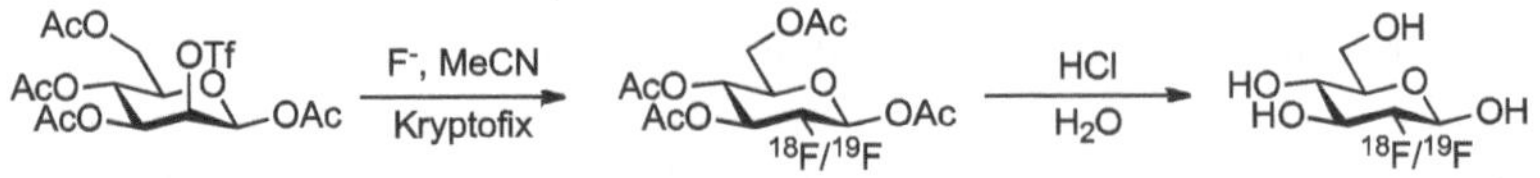

Scheme 4. Synthesis of the radiolabeled imaging probe [^{18}F]FDG 28 in a PDMS-based microfluidic reactor

positron emission tomography (PET) imaging studies in mice (Lee et al. 2005).

10 Microreactor Synthesis of β-Peptides

β-Amino acid oligomers (β-peptides) present a unique class of peptides. In contrast to their natural α-amino acid counterparts, β-peptides require relatively few residues to form secondary structures like turns, helices, or sheets (Seebach et al. 1996, 2001, 2004; Cheng et al. 2001). However, these secondary structures can also severely hamper the synthesis of these compounds (Murray and Gellman 2005; Lelais et al. 2006). To increase the synthetic efficiency of helical β-peptides, microwave-assisted solid phase peptide synthesis (SPPS) protocols have also been applied (Murray and Gellman 2005; Lelais et al. 2006; Erdélyi and Gogoll 2002).

Recently, we described the application of a silicon microfluidic reactor to the assembly of oligo-β-peptides (Flögel et al. 2006). The microreactor not only allowed for quick scanning of reaction conditions, but also the procurement of synthetically useful amounts of peptides (Curran 2001; Curran and Luo 1999; Zhang 2004).

The assembly of tetrapeptide **19** that contains all possible β-dipeptide bonds, (β^3-β^3)-, (β^3-β^2)-, and (β^2-β^3), and also a turn inducing β^3-(*R*)-Ala-β^2-(*R*)-Val element was achieved employing a Boc-strategy (Scheme 5). A fluorous benzyl group was incorporated in the first amino acid to streamline the purification procedure by fluorous solid phase extraction (FSPE) (Filippov et al. 2002; de Visser et al. 2003). Thus, the assembly of the fully protected tetrapeptide commenced with the construction of the first β^3-β^3-peptide bond by applying the previously established conditions. A residence time of 3 min at 90°C provided the Boc-protected dipeptide **15** in 91% isolated yield after FSPE. Notably, the product precipitated in the collection flask, which was kept at ambient temperature, indicating the poor solubility of this class of compounds (Hessel et al. 2005).

For the next condensation, the reaction progress was scanned at both 90°C and 120°C, since it was anticipated that lower reaction temperatures would present insurmountable solubility problems. The construc-

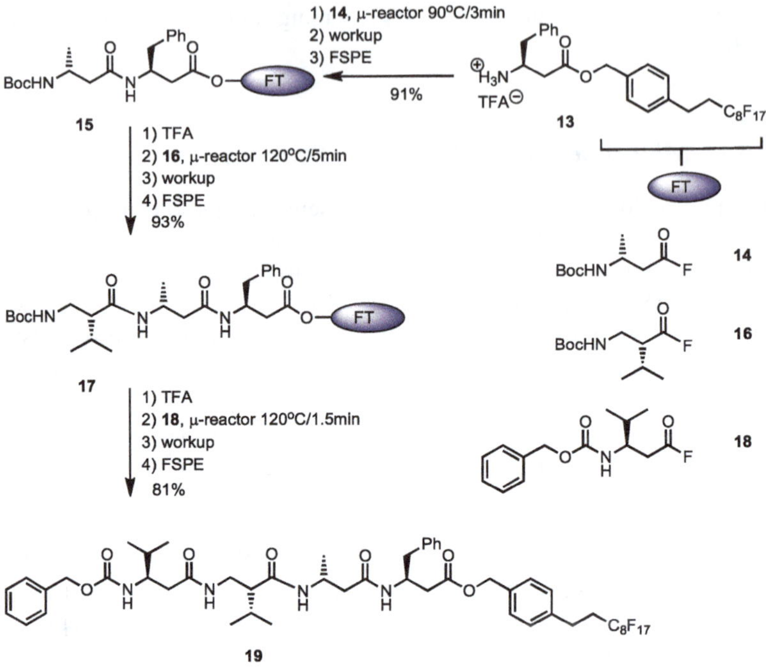

Scheme 5. Assembly of tetrapeptide **19**

tion of the more sterically demanding β^3-β^2 peptide bond required higher temperatures and/or longer residence times. Maximum conversion was obtained by running the reaction at 120°C for 5 min. It should be noted that the low solubility of the product prohibited longer reaction times. Purification of the product was readily accomplished by FSPE to provide tripeptide **17** in 93% yield on 0.4-mmol scale. The final β^2- β^3 condensation was executed at 120°C and a reaction scan revealed the coupling to be complete after 1.5 min. Tetramer **19** was synthesized on 0.2-mmol scale and obtained in 81% yield. Cleavage of the fluorous tag and removal of the benzyloxycarbonyl function provided the zwitter ionic target peptide.

To compare the here described methodology to traditional assembly strategies, the tetrapeptide was also synthesized on solid support (Fmoc strategy, HATU-D*i*PEA-mediated peptide couplings), and in solution using conventional laboratory glassware. All procedures provided the tetramer in comparable overall yield. However, purification of the fluorous peptides was exceedingly facile as compared to the purification of the nonfluorous peptides. In addition, it should be noted that the microreactor system allowed the precise control over unconventionally high reaction temperatures, thereby significantly reducing reaction time and preventing the precipitation of the products during the reaction, which led to inhomogeneous gel-like reaction mixtures when using traditional laboratory glassware.

Not only can the synthetic strategy find applications in the construction of challenging peptides, it can also be applied for the assembly of other biopolymers, such as oligosaccharides and nucleic acids.

11 Conclusion and Outlook

Many chemical reactions can become faster, safer, and cleaner. The down-scaling of reaction volumes in microreactors offers a means to better control reaction conditions, including temperature, time, mixing, and to use minute amounts of precious compounds to rapidly screen a variety of conditions, generating a wealth of information on reaction kinetics and pathways. Microreactors present opportunities to apply conditions that are inaccessible using conventional laboratory equipment, such as superheated solvents, and reactions in explosive regimes. However, there are also inherent drawbacks associated with the miniaturized format: the reactors are incompatible with solid reagents, very sensitive to precipitating products, and are synthetically mainly useful for relatively fast reactions (in turn, slow reactions can be turned into fast ones using unconventional conditions only achievable in microreactors!). The efficient and effective analysis of reaction mixtures in a high-throughput format is a major issue. Real-time on-line analysis has been accomplished using mass spectrometry, IR spectroscopy and UV-VIS absorbance, and currently on-chip NMR technology is being developed. However, organic synthesis of complex molecules requires analytical methods that can distinguish between regio- and stereochemically different compounds, for which MS, IR, UV-VIS, and low-resolution NMR are inadequate. Off-line HPLC, LCMS, or GC-MS are currently the most commonly applied techniques to analyze reactions conducted in microreactors.

Before microreactor technology will be employed as a standard research tool, it will have to become more readily available commercially. Eventually, cheap, easy-to-use, flexible microreactors will be used as a valuable alternative to the round-bottomed flask.

References

Acke DRJ, Orru RVA, Stevens CV (2006) Continuous synthesis of tri- and tetra-substituted imidazoles via a multicomponent reaction under microreactor conditions. QSAR Comb Sci 25:474–483

Antes J, Tuercke T, Marioth E, Lechner F, Scholz M, Schnürer F, Krause HH, Löbbecke S (2001) In: Matlosz M, Ehrfeld W, Baselt JP (eds) IMRET 5: Proceedings of the Fifth International Conference on Microreaction Technology. Springer Berlin New York Heidelberg, pp 446–453

Barrow D, Cefai J, Taylor S (1999) Shrinking to fit. Chem Ind 15:591–594

Bradley D (1999) Chemical Reduction. Eur Chem 1:17–20

Chambers RD, Spink RCH (1999) Microreactors for Elemental Flourine. Chem Comm 10:883–884

Chambers RD, Holling D, Spink RCH, Sandford G (2001) Elemental Fluorine – Part 13. Gas-Liquid Thin Film Microreactors for Selective Direct Fluorination. Lab Chip 1:132–137

Chambers RD, Fox MA, Sandford G (2005) Elemental fluorine – Part 18. Selective Direct Fluorination of 1,3-Ketoesters and 1,3-Diketones using Gas/Liquid Microreactor Technology. Lab Chip 5:1132–1139

Cheng RP, Gellman SH, DeGrado WF (2001) Beta-Peptides: From Structure to Function. Chem Rev 101:3219–3232

Cooper J, Disley D, Cass T (2001) Microsystems Special – Lab-On-A-Chip and Microarrays. Chem Ind 20:653–655

Cowen S (1999) Chip service. Chem Ind 15:584–586

CPC (2007) Cytos Lab System. http://www.cpc-net.com/cytosls.shtml. Cited 5 February 2007

Curran DP (2002) Fluorous Reverse Phase Silica Gel. A New Tool For Preparative Separations in Synthetic Organic and Organofluorine Chemistry. Synlett 1488–1496

Curran DP, Luo ZY (1999) Fluorous Synthesis with Fewer Fluorines (Light Fluorous Synthesis): Separation of Tagged from Untagged Products by Solid-Phase Extraction with Fluorous Reverse-Phase Silica Gel. J Am Chem Soc 121:9069–9072

De Mas N, Jackman RJ, Schmidt MA, Jensen KF (2002) In: Matkisz M, Ehrfeld W, Baselt JP (eds) IMRET 5: Proceedings of the Fifth International Conference on Microreaction Technology. Springer Berlin, New York, Heidelberg, pp 60–67

De Mas N, Günther A, Schmidt MA, Jensen KF (2003) Microfabricated Multiphase Reactors for the Selective Direct Fluorination of Aromatics. Ind Eng Chem Res 42:698–710

De Visser PC, van Helden M, Filippov DV, van der Marel GA, Drijfhout JW, van Boom JH, Noort D, Overkleeft HS (2003) A novel, Base-Labile Fluorous Amine Protecting Group: Synthesis and Use as a Tag in the Purification of Synthetic Peptides. Tetrahedron Lett 44:9013–9016

DeWitt SH (1999) Microreactors for Chemical Synthesis. Curr Opin Chem Biol 3:350–356

Doku GN, Haswell SJ, McCreedy T, Greenway GM (2001) Electric Field-Induced Mobilisation of Multiphase Solution Systems Based on the Nitration of Benzene in a Microreactor. Analyst 126:14–20

Duan J, Sun L, Liang Z, Zhang J, Wang H, Zhang L, Zhang W, Zhang Y (2006) Rapid Protein Digestion and Identification using Monolithic Enzymatic Microreactor Coupled with Nano-Liquid Chromatography-Electrospray Ionization Mass Spectrometry. J Chromatogr A 1–2:165–174

Ehrfeld Mikrotechnik BTS (2007) http://www.ehrfeld-shop.biz/shop/catalog/index.php. Cited 5 February 2007

Ehrfeld W, Hessel V, Löwe H (2000) Microreactors: new technology for modern chemistry. Wiley-VCH, Weinheim

Ehrich H, Linke C, Morgenschweis K, Baerns M, Jahnisch K (2002) Application of Microstructured Reactor Technology for the Photochemical Chlorination of Alkylaromatics. Chimia 56:647–653

Filippov DV, van Zoelen DJ, Oldfield SP, van der Marel GA, Overkleeft HS, Drijfhout JW, van Boom JH (2002) Use of Benzyloxycarbonyl (Z)-Based Fluorophilic Tagging Reagents in the Purification of Synthetic Peptides. Tetrahedron Lett 43:7809–7812

Erdélyi M, Gogoll A (2002) Rapid Microwave-Assisted Solid Phase Peptide Synthesis. Synthesis 1592–1596

Fletcher PDI, Haswell SJ (1999) Downsizing Synthesis. Chem Ber 35:38–41

Fletcher PDI, Haswell SJ, Paunov VN (1999) Theoretical Considerations of Chemical Reactions in Microreactors Operating under Electroosmotic and Electrophoretic Control. Analyst 124:1273–1282

Fletcher PDI, Haswell SJ, Pombo-Villar E, Warrington BH, Watts P, Wong SYF, Zhang XL (2002) Microreactors: Principles and Applications in Organic Synthesis. Tetrahedron 58:4735–4757

Flögel O, Codée JDC, Seebach D, Seeberger PH (2006) Microreactor Synthesis of Beta-Peptides. Angew Chem Int Ed 45:7000–7003

Geyer K, Codee JDC, Seeberger PH (2006) Microreactors as Tools for Synthetic Chemists – The Chemists' Round-Bottomed Flask of the 21st Century? Chem Eur J 12:8434–8442

Geyer K, Seeberger PH (2007) Optimization of Glycosylation Reactions in a Microreactor. Helv Chim Acta 90:395–403

Greenway GM, Haswell SJ, Morgan DO, Skelton V, Styring P (2000) The Use of a Novel Microreactor for High Throughput Continuous Flow Organic Synthesis. Sens Actuator B Chem 63:153–158

Hansen CL, Classen S, Berger JM, Quake SR (2006) A Microfluidic Device for Kinetic Optimization of Protein Crystallization and In Situ Structure Determination. J Am Chem Soc 128:3142–3143

Haswell SJ, Middleton RJ, O'Sullivan B, Skelton V, Watts P, Styring P (2001a) The Application of Microreactors to Synthetic Chemistry. Chem Commun 5:391–398

Haswell SJ, O'Sullivan B, Styring P (2001b) Kumada-Corriu Reactions in a Pressure-Driven Microflow Reactor. Lab Chip 1:164–166

Hessel V, Ehrfeld W, Golbig K, Haverkamp V, Löwe H, Storz M, Wille C, Guber A, Jahnisch K, Baerns M (2000) In: Ehrfeld W (ed) IMRET 3: Proceeding of the Third International Conference on Microreaction Technology. Springer Berlin, New York, Heidelberg, pp 526–540

Hessel V, Kolb G, de Bellefon C (2005a) Catalytic Microstructured Reactors – Preface. Catal Today 110:1–1

Hessel V, Löwe H, Müller A, Kolb G (2005b) Chemical micro process engineering 1 + 2. Fundamentals modelling and reactions/processes and plants. Wiley-VCH, Weinheim

Jähnisch K, Baerns M, Hessel V, Ehrfeld W, Haverkamp V, Löwe H, Wille C, Guber A (2000) Direct Fluorination of Toluene Using Elemental Fluorine in Gas/Liquid Microreactors. J Fluor Chem 105:117–128

Jähnisch K, Hessel V, Löwe H, Baerns M (2004) Chemistry in Microstructured Reactors. Angew Chem Int Ed 43:406–446

Jensen KF (2001) Microreaction Engineering – is Small Better? Chem Eng Sci 56:293–303

Kawaguchi T, Miyata H, Ataka K, Mae K, Yoshida J (2005) Room-Temperature Swern Oxidations by Using a Microscale Flow System. Angew Chem Int Ed 44:2413–2416

Kiwi-Minsker L, Renken A (2005) Microstructured reactors for catalytic reactions. Catal Today 110:2–14

Kockmann N, Brand O, Fedder GK (2006) Micro process engineering. Wiley-VCH, Weinheim

Kolb G, Hessel V (2004) Micro-Structured Reactors for Gas Phase Reactions. Chem Eng J 98:1–38

Lee CC, Sui GD, Elizarov A, Shu CYJ, Shin YS, Dooley AN, Huang J, Daridon A, Wyatt P, Stout D, Kolb HC, Witte ON, Satyamurthy N, Heath JR, Phelps ME, Quake SR, Tseng HR (2005) Multistep Synthesis of a Radiolabeled Imaging Probe Using Integrated Microfluidics. Science 310:1793–1796

Lelais G, Seebach D, Jaun B, Mathad RI, Flögel O, Rossi F, Campo M, Wortmann A (2006) Beta-Peptidic Secondary Structures fortified and enforced by Zn^{2+} Complexation – On the Way to Beta-Peptidic Zinc Fingers? Helv Chim Acta 89:361–403

Lu H, Schmidt MA, Jensen KF (2001) Photochemical Reactions and On-Line UV Detection in Microfabricated Reactors. Lab Chip 1:22–28

Manz A, Harrison DJ, Verpoorte EMJ, Fettinger JC, Ludi H, Widmer HM (1991) Miniaturization of Chemical-Analysis Systems – A Look into next Century Technology or just a Fashionable Craze. Chimia 45:103–105

McCreedy T (1999) Reducing the Risks of Synthesis. Chem Ind 15:588–590

McCreedy T (2000) Fabrication Techniques and Materials Commonly Used for the Production of Microreactors and Micro Total Analytical Systems. Trac Trends Anal Chem 19:396–401

McCreedy T (2001) Rapid Prototyping of Glass and PDMS Microstructures for Micro Total Analytical Systems and Micro Chemical Reactors by Microfabrication in the General Laboratory. Anal Chim Acta 427:39–43

Murray JK, Gellman SH (2005) Application of Microwave Irradiation to the Synthesis of 14-Helical Beta-Peptides. Org Lett 7:1517–1520

Nikbin N, Watts P (2004) Solid-Supported Continuous Flow Synthesis in Microreactors Using Electroosmotic Flow. Org Process Res Dev 8:942–944

Panke G, Schwalbe T, Stirner W, Taghavi-Moghadam S, Wille S (2003) A Practical Approach of Continuous Processing to High Energetic Nitration Reactions in Microreactors. Synthesis 18:2827–2830

Pennemann H, Watts P, Haswell SJ, Hessel V, Löwe H (2004) Benchmarking of Microreactor Applications. Org Process Res Dev 8:422–439

Ratner DM, Murphy ER, Jhunjhunwala M, Snyder DF, Jensen KF, Seeberger PH (2005) Microreactor-Based Reaction Optimization in Organic Chemistry – Glycosylation as a Challenge. Chem Comm 5:578–580

Rouhi AM (2004) Microreactors Eyed for Industrial Use. Chem Eng News 82:18–19

Saaby S, Knudsen KR, Ladlow M, Ley SV (2005) The Use of a Continuous Flow-Reactor Employing a Mixed Hydrogen-Liquid Flow Stream for the Efficient Reduction of Imines to Amines. Chem Commun 23:2909–2911

Seebach D, Overhand M, Kühnle FNM, Martinoni D, Oberer L, Hommel U, Widmer H (1996) Beta-Peptides: Synthesis by Arndt-Eistert Homologation with Concomitant Peptide Coupling. Structure Determination by NMR and CD Spectroscopy and by X-ray Crystallography. Helical Secondary Structure of a Beta-Hexapeptide in Solution and its Stability Towards Pepsin. Helv Chim Acta 79:913–941

Seebach D, Schreiber JV, Arvidsson PI, Frackenpohl J (2001) The Miraculous CD Spectra (and Secondary Structures?) of Beta-Peptides as they Grow Longer – Preliminary Communication. Helv Chim Acta 84:271–279

Seebach D, Beck AK, Bierbaum DJ (2004) The World of Beta- and Gamma-Peptides Comprised of Homologated Proteinogenic Amino Acids and Other Components. Chem Biodiv 1:1111–1239

Skelton V, Greenway GM Haswell SJ, Styring P, Morgan DO, Warrington B, Wong SYF (2001a) The preparation of a Series of Nitrostilbene Ester Compounds Using Microreactor Technology. Analyst 126:7–10

Skelton V, Greenway GM, Haswell SJ, Styring P, Morgan DO, Warrington BH, Wong SYF (2001b) The Generation of Concentration Gradients Using Electroosmotic Flow in Microreactors Allowing Stereoselective Chemical Synthesis. Analyst 126:11–13

Snyder DA, Noti C, Seeberger PH, Schael F, Bieber T, Rimmel G, Ehrfeld W (2005) Modular Microreaction Systems for Homogeneously and Heterogeneously Catalyzed Chemical Synthesis. Helv Chim Acta 88:1–9

Syrris (2007) http://www.syrris.com. Royston, UK. Cited 5 February 2007

Taghavi-Moghadam S, Kleemann A, Overbeck S (2000) VDE World Microtechnologies Congress MICROtech 2000, Vol. 2, VDE, Berlin, p 489

Thales Nano (2006) http://thalesnano.com/. Cited 5 February 2007

Thayer AM (2005) Harnessing Microreactions. Chem Eng News 83:43–52

Wang B, Zhao Q, Wang F, Gao C (2006) Biologically Driven Assembly of Polyelectrolyte Microcapsule Patterns To Fabricate Microreactor Arrays. Angew Chem Int Ed 45:1560–1563

Watts P, Haswell SJ (2005) The Application of Microreactors for Organic Synthesis. Chem Soc Rev 34:235–246

Watts P, Wiles C, Haswell SJ, Pombo-Villar E, Styring P (2001) The Synthesis of Peptides Using Microreactors. Chem Comm 11:990–991

Watts P, Wiles C, Haswell S, Pombo-Villar E (2002a) Investigation of Racemisation in Peptide Synthesis within a Microreactor. Lab Chip 2:141–144

Watts P, Wiles C, Haswell S, Pombo-Villar E (2002b) Solution Phase Synthesis of Beta-Peptides Using Microreactors. Tetrahedron 58:5427–5439

Wehle D, Dejmek M, Rosenthal J, Ernst H, Kampmann D, Trautschold S, Pechatschek R (2000) DE10036603 A

Wiles C, Watts P, Wiles C, Haswell S, Pombo-Villar E (2001) The Aldol Reaction of Silyl Enol Ethers within a Microreactor. Lab Chip 1:100–101

Zhang W (2004) Fluorous Synthesis of Heterocyclic Systems. Chem Rev 104:2531–2556

Ernst Schering Foundation Symposium Proceedings, Vol. 3, pp. 21–37
DOI 10.1007/2789_2007_026

Published Online: 23 May 2007

Micro-Fluidic and Lab-on-a-Chip Technology

X. Zhang, S.J. Haswell(✉)

Department of Chemistry, The University of Hull, HU6 7RX Hull, UK
email: *S.J.Haswell@Hull.ac.uk*

Abstract. By reducing the operational dimensions of a conventional macro-fluidic-based system down to the micron scale, one can not only reduce the sample volume, but also access a range of unique characteristics, which are not achievable in conventional macro-scale systems. This chapter will discuss the unique properties of miniaturised systems based on micro-fluidic and Lab-on-a-Chip technology and consider how these may influence the overall performance associated with chemical and biological processing. Some consideration will also be given to the selection of materials and/or surface modifications that will be proactive in exploiting the high surface area and thermal and mass transfer properties, to enhance process performance.

1 Introduction

The development of miniaturised systems based on micro-fluidic and Lab-on-a-Chip technology has witnessed an explosive growth over the last decade (Manz and Becker 1998; Jensen 2001; Fletcher et al. 2002a; Andersson and van den Berg 2004; Whitesides 2006). This can largely be attributed to the unique properties of such micro-systems, which represent the potential to "shrink" conventional bench chemical systems to the size of a few square centimetres with major advantages in terms of speed, performance, integration, portability, sample/solvent quantity, automation, hazard control and cost. These advantages are important for a variety of applications in analytical chemistry, biochemistry, clinical diagnosis, medical chemistry and industrial chemistry (Laurell et al. 2004; Haber 2006). Figure 1 shows a range of such devices designed and fabricated from different materials at Hull University.

The main feature of a micro-fluidic-based system is the presence of a micron-scale channel network in which chemical reactions and/or bio-

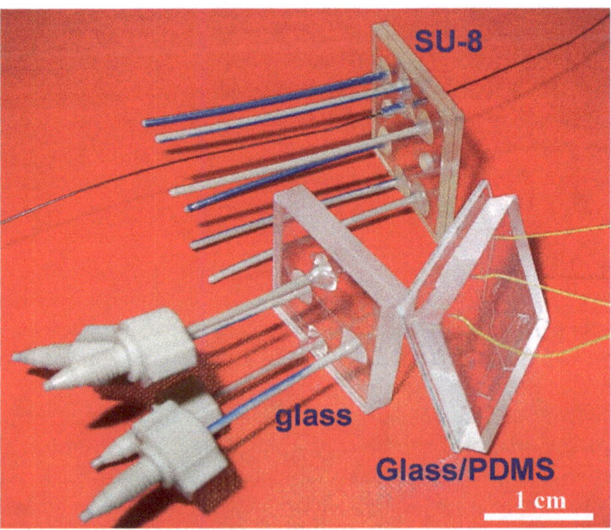

Fig. 1. Micro-fluidic chips made of different material for chemical synthesis and biochemical and chemical processing and analysis

logical systems are brought together, using a variety of pumping techniques, for reaction, separation or analysis. This scaling-down of the operation space dimension (compared to the conventional macro-scale system) not only reduces the sample volume, but also brings unique characteristics into the micro-scale fluidic environment. These unique operating characteristics include (a) a laminar flow environment where diffusion dominates mass transfer, (b) a high surface area-to-volume ratio, which enhances surface effects including rapid heat and mass transfer, (c) spatial and temporal control of reagents and products including the presence of localised concentration gradients, and (d) the opportunity to integrate processes and measurement systems on a single technology platform leading to the concept of the Lab-on-a-Chip. In this chapter, the fundamental and practical advantages associated with micro-fluidic and Lab-on-a-Chip technology are discussed with reference to specific examples in order to illustrate the unique characteristics of such miniaturised Lab-on-a-Chip systems.

2 Laminar Flow with Diffusion-Dominated Mass Transfer

It is well known that the flow within micro-channels lies in the laminar flow regime with a typical Reynolds number (representing the ratio of inertia forces to viscous forces, a measure to determine whether a flow is laminar or turbulent) below 10, compared to a Reynolds number greater than 3,000, which is normally associated with turbulent flow in pipes. The simplified flow conditions of a laminar flow system enable a relatively accurate prediction to be made of the flow behaviour, which consequently leads to the highly controlled manipulation of flows within a micro-fluidic channel network. Figure 2 shows a laminar flow within a T-shaped channel network where a stream of dye solution and a stream of solvent meet and flow side by side along the main channel. It can be clearly seen that at the interface of the two streams, diffusion occurs in a way that can be predicted in terms of either time or position due to the flow regime. By using an absorbance measurement technique based on the Lambert-Beer law, the concentration profiles with different mixing times can be obtained (Fletcher et al. 2002b). The theoretical prediction

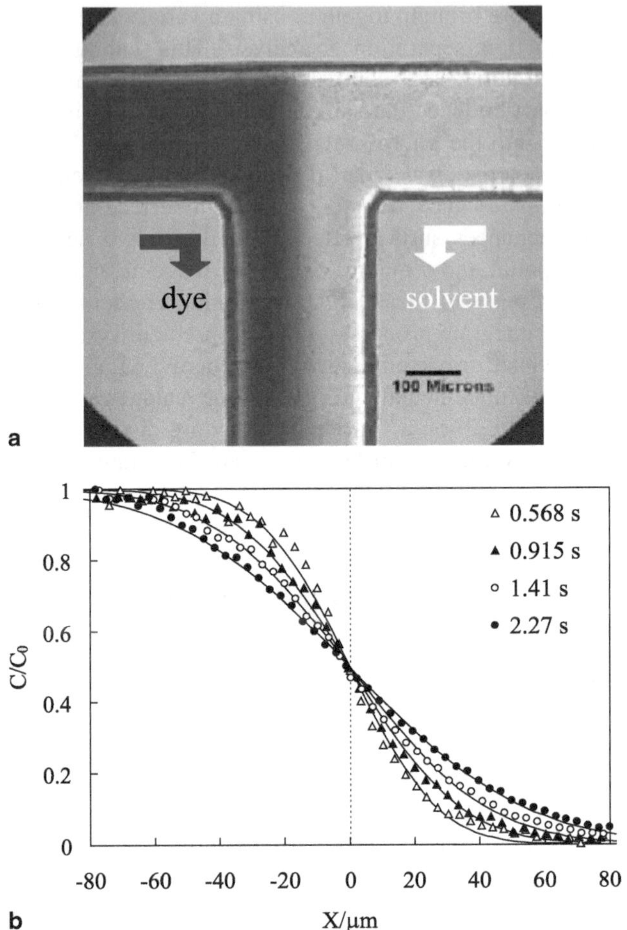

Fig. 2. a Laminar flow within a T-shaped channel network and **b** the concentration profiles at different diffusive mixing times corresponding to different positions along the main channel

of the concentration change is also plotted on Fig. 2b, showing good agreement with the experimental results.

Under laminar flow conditions, the mass transfer will be dominated by diffusion. Based on Fick's law (Fick 1855), the relationship between the travel distance (L) of a molecule by diffusion and time (t) can be simplified as:

$$L = (2D \cdot t)^{1/2} \quad (1)$$

where D is the diffusion coefficient.

It can be seen from Eq. 1 that by scaling down the dimension in which diffusive mixing occurs, a significant reduction in the time taken to achieve complete mixing is achieved. For example, a water molecule takes 200 s to diffuse across a 1-mm-wide channel, but only takes 500 ms to cross a 50-μm-wide channel (where the self-diffusion coefficient of water at 25°C is 2.30×10^{-9} m^2 s^{-1}) (Mills 1973). This significant reduction in mixing time is beneficial for controlling reaction progress, in particular for initiating or quenching reactions in a controlled manner, enabling improvements to be made in product selectivity.

3 High Surface Area-to-Volume Ratio

When scaling a conventional centimetre-sized reactor down to the micron scale, the surface-to-volume ratio significantly increases to the point where the container walls can effectively become an active or influential part of the reaction or process occurring in the fluidic channel. Clearly this attribute of micro-reactors can be viewed in a positive way and leads to the opportunity of exploiting surface-dependent performance including electrokinetic driven flow, surface functionalisation and mass transfer, and heat transfer.

3.1 Electro-osmotic Flow and Electrophoresis

A relatively simple, but important example, of the effect of high surface-to-volume ratio is where the surface charge of the capillary is neutralised by the solution contained within it to form a charged double layer, which under the influence of an applied electric field leads to the electro-osmotic mobilisation of the solution. This type of pumping method based on electro-osmotic flow (EOF), using voltages applied via electrodes placed at the ends of a capillary channel, in say a reservoir, has several significant advantages over alternative pumping methods (Fletcher et al. 2002a,b). It can be easily miniaturised because no mechanical moving parts are involved and the required voltage sequences can be readily applied under automated computer control. For glass and many polymer-based micro-fluidic devices, the channel wall–solution interface normally has a negative charge. This immobile surface charge attracts a diffuse layer (of thickness of the order of nanometers) of mobile, oppositely charged counterions in the solution adjacent to the channel wall (cations for a negatively charged glass channel wall). Upon the application of an electric field along the channel length, the nanometer-thick skin of mobile cations migrates towards the more negative electrode and drags all the intervening solution in the bulk of the channel with it (Overbeek 1952; Rice and Whitehead 1965; Hunter 1981). An important feature of EOF is that the liquid EOF velocity is constant across the channel except in the nanometer-thick regions of the diffuse layer of counterions very close to the wall. Unlike EOF, which has plug flow, pressure-driven flow produces a parabolic velocity profile with high velocities in the channel centre and slow velocities near the wall, giving rise to increased blurring of reagent zones along a channel length.

In addition to EOF, chemical species within the electric field also have an additional electrophoretic velocity. The magnitude of electrophoretic velocity for typical micro-reactor operating voltages is commonly comparable with EOF velocity. Thus, the total velocity of a charged species is given by the vector sum of EOF and electrophoretic velocities. The direction of EOF for aqueous solutions in a glass micro-reactor is normally towards the more negative electrode. Cationic and neutral charged species will also migrate in the same direction as EOF (i.e. towards the more negative electrode), with cations moving faster down the channel than the solvent and neutral species. Anionic solutes will be retarded and may, if the magnitude of electrophoretic mobility is greater than EOF, move in the opposite direction. In this way, electrophoretic separation of solutes occurs along with EOF in the micro-reactor channels when operated under electrokinetic control. The ability to use electrophoretic mobility to spatially locate charged reagents and products within a micro-fluidic reactor independently of the solvent forms a useful aspect of reaction control (Fletcher et al. 2002b). Figure 3 shows schematically the voltage-driven flow within a micro-channel with both EOF and electrophoretic movement where, depend-

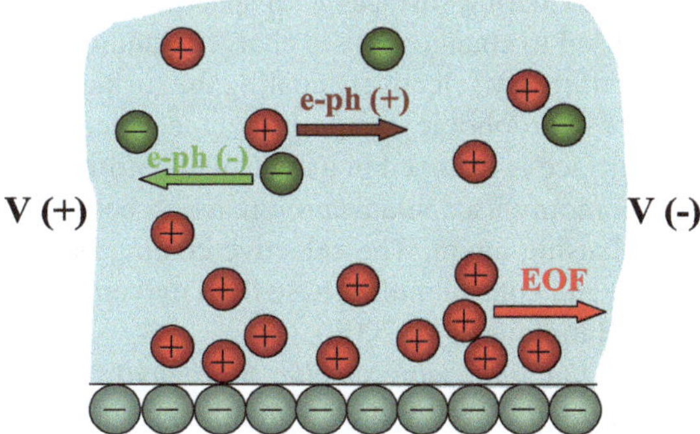

Fig. 3. Voltage-driven movement of the diffuse layer of cations adsorbed at the negatively charged channel wall produces an EOF and mobility of differently charged species

ing on species' charge, the electrophoretic velocity is either accelerated or retarded (or even reversed in direction) relative to EOF.

3.2 Surface Functionalisation

The relatively large surface area within micro-fluidic structures can be exploited in a proactive way to add chemical and or biological functionalisation to a process. A range of surface modification techniques have been developed and some of the more widely used techniques are outlined below.

1. *To alter the electric charge on surfaces.* This is normally carried out by either plasma treatment or coating of surfaces with silanising agents. By changing the surface charge the EOF direction is altered to control the sample delivery within the channel network. Furthermore, patterning the channel surface with different charges can result in some counter-conflict flows within the channel, which can significantly improve mixing (Hau et al. 2003). Under different plasma treatment conditions, the surface charge (Hu et al. 2002) and/or the contact angle can be changed, either temporarily or permanently (Gillmor et al. 2002; Fritz and

Owen 1995). Coating surfaces with specific chemical groups has also been used to change surface charge (Handique et al. 2000).

2. *Surface wetting*. By selectively coating the surface with either hydrophilic or hydrophobic groups the contact behaviour of reagents with the surface is changed in a controllable format. This is done by the commonly used silanisation reaction between the surface and the silansing agent. The selective coating has been applied to generate and control multi-phase fluid movement along micro-channels (Handique et al. 2000).
3. *Surface modification with specific biological functions to bind different biological molecules*. A multi-step surface treatment is generally required. Firstly, a layer of chemical coating such as amino groups is created. This is then bio-functionalised by binding biological molecules, for instance, antibodies. Finally, uncoated sites are blocked to prevent nonspecific adsorption. This technique has been extensively used in immunoassay (Lagally and Mathies 2004) and DNA arrays on microslides (Schneider et al. 2001). Based on the same principle, the surface within micro-channels can be modified *in situ*, and this technique is increasingly attracting attention following the pioneering work by Whitesides et al. (Takayama et al. 1999). Early results indicate that the micro-fluidic bio-modification can promise a continuous, faster and flexible tool for bioanalysis.
4. *Adding additional surface*. Whilst a high ratio of surface to volume is inherent in an empty/open micro-channel, this characteristic may be further pronounced by packing micro-beads in the channel (He et al. 2004). The introduction of beads within micro-channels can introduce some practical difficulties, but a number of successful applications using this type of methodology have been described (He et al. 2004; Nikbin and Watts 2004).

The use of silica and polymer monoliths (Svec 2004; Peterson et al. 2003; Yang et al. 2005) can also significantly increase the surface area-to-volume ratio. In addition, the porous monoliths can be equipped to comprise reactive functionalities or even reactants to perform specific chemical or biochemical reactions (Svec 2004). Two main problems, however, may be encountered during the operation of monoliths: (a) the shortcut flow may exist and (b) the surface may not be completely functionalised with reagents. These problems are mainly due to the unevenly distributed structures within the porous monoliths and can be overcome by using *in situ* polymerisation where the functionalised monomer is introduced into the micro-channel and then polymerised subject to curing conditions, for example, an exposure to UV light Peterson et al. 2003; Yang et al. 2004).

It should be noted that rather than exploiting the proactive aspects of a surface, it is equally valid to mask or negate wall effects by fluidically isolating the sample from the substrate channel wall to eliminate surface effects. This is carried out by using either a co-axial flow to keep the sample in the centre, i.e. away from the walls (Takagi et al. 2004) or in a similar way using multiple flow streams to surround the sample stream to form a sheath (Munson et al. 2004, 2005).

3.3 Enhanced Heat and Mass Transfer

The high surface-to-volume ratio can also significantly improve both thermal and mass transfer conditions within micro-channels in two ways; firstly, the convective heat and mass transfers, which take place at the multi-phase interface, are improved via a significant increase in heat and mass transfer area per unit volume. Secondly, heat and mass transfers within a small volume of fluid take a relatively short time to occur, enabling a thermally and diffusively homogeneous state to be reached quickly. The improvement in heat and mass transfer can certainly influence overall reaction rates and, in some cases, product selectivity. Perhaps one of the more profound effects of the efficient heat and mass transfer property of micro-reactors is the ability to carry potentially explosive or highly exothermic reactions in a safe way, due to the relatively small thermal mass and rapid dissipation of heat.

On the other hand, the enhanced heat dissipation from the small volume of reagent into the relatively large volume of reactor body can be disadvantageous where the sample within micro-channels has to be heated up. In a conventional heating method, for example using a thermal bath or oven, the reactor body needs to be heated up first and then the heat is transferred by thermal conduction into the sample in the micro-channels. In this case, only a very small portion of heat is effectively applied to the sample and, more importantly, it will be a time-consuming process, even impossible if a quick switch of temperature or temperature cycling is required.

As a remote heating source microwave radiation has been used widely for domestic (e.g. microwave oven), industrial (e.g. microwave drying) (Metaxas and Meredith 1983), and scientific research (e.g. microwave reaction chemistry) (Zhang and Hayward 2006). In its application to Lab-on-a-Chip systems, microwave heating offers opportunities to overcome the difficulties discussed above. Firstly, volumetric heating of microwave radiation provides a fast heating method so the sample can reach a high heating rate. Secondly, microwave energy can be absorbed selectively by different material depending on its dielectric property, with higher dielectric loss leading to more energy being absorbed. To enhance the radiation absorbance, a stronger microwave absorber can be employed. Since glass is generally regarded as microwave transparent microwave radiation can be mainly absorbed by the sample with the micro-channel and the added absorber. Thirdly, microwave heating can be switched on/off in almost a simultaneous way, so the sample temperature can be changed quickly owing to its relatively large surface area. Taking all these into account, a microwave heating micro-reactor system has been developed for heterogeneous catalysis study. As shown in Fig. 4, a glass-made micro-reactor with a packing catalytic bed is placed in a microwave cavity where the electromagnetic field can be program-controlled in terms of strength and time for temperature control. On the chip bottom in a specific area, underneath the catalyst bed in this case, a patch of golden film (10 nm thick) is coated (Fig. 4b). It has been found that this strong microwave absorber (golden film) can increase the heating rate significantly (He et al. 2004).

4 Spatial and Temporal Evolution of Reactions

Under the diffusive laminar flow conditions, the ability to add reagents at specific locations or times leads to the unique ability to control and monitor the spatial and temporal domain of dynamic chemical processes. This attribute has some analogies with the control exerted on biochemical reactions by the micron-scale structures of living cells. Ex-

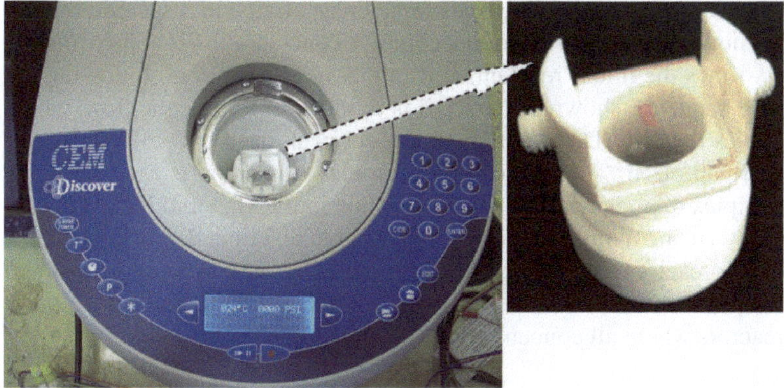

Fig. 4. Microwave heating system showing a micro-fluidic chip with a golden film (10 nm thick) patterned on the channel and to act as an efficient heating patch, which will have high thermal energy but low thermal mass

ploitation of this effect, such that a reaction well occurs in a position where the local concentration of a key intermediate is high, is a potentially valuable approach towards controlling the yield and selectivity of reactions.

Figure 5 shows an example of this type of mixing, reaction and separation under electrokinetic-driven flow conditions within a double-T shaped microchannel network (Fletcher et al. 2002b). In this process, a slug of the uncharged ligand, pyridine-azo-dimethylaniline (PADA), is injected into a flowing stream of Ni^{2+} ions. Upon applying an electric field along the main channel, the slug of PADA becomes displaced by the stream of Ni^{2+} ions because of its electrophoretic mobility. In this case, mixing (and hence reaction) occurs only by interdiffusion at the leading and trailing edges of the slug of PADA. For the wide slug and the short time shown, the interdiffusion and therefore the extent of the reaction is low. The time for diffusive mixing across the slug can be made shorter by reducing the slug width. To maximise the conversion to products in this mixing mode, a series of many narrow slugs are therefore more effective than a single broad slug. It should be noted that, in contrast to diffusive mixing, mixing by differential electrophoretic mobilities does not result in dilution of the localised reagent concentrations, enabling charged species to be concentrated in a specific region; as shown in Fig. 5b the product complex concentration is approximately six times the initial reactant concentration. Furthermore, following the time sequence shown in Fig. 5, it is possible to reverse the flow, which moves the $NiPADA^{2+}$ complex product peak back into the concentration gap of Ni^{2+}. As the complex formation reaction is reversible, the complex dissociates via the back reaction into PADA and Ni^{2+}. This example, using a mechanistically very simple reaction, demonstrates how micro-reactors can be used to realise an extra dimension of spatial and temporal control of reactions, which is unachievable in bulk solution reactors where all concentrations are uniform.

5 Integration of Processes and Measurement Systems and Their Automation

Each of the properties of a micro-reactor outlined above does not have to be exploited independently but can be combined to provide multiple functionality within one micro-reactor. In this way, multi-step processes, combining a range of physical and chemical steps, can be performed in a controlled and reproducible way. In addition, the integration of *in situ*, real time or end-of-line analytical measurements can be effectively realised using micro-reactor technology, leading to rapid automated methodology. A combination of these two features will clearly create tools for the pharmaceutical and fine chemical industries, where high throughput and information-rich techniques are constantly sought for the rapid evaluation of reaction arrays. Apart from the greatly reduced reaction times associated with micro-reactors, handling times to assay and chemical reagent costs may be virtually eliminated. This paradigm is shown diagrammatically in Fig. 6 (Fletcher et al. 2004).

The concept of an integrated micro-fluidic-based system has now been developed (Zhang et al. 2006), with an example shown in Fig. 7. This particular system is based on conventional chromatographic instrumentation and employs a multi-valving system, located between two syringe pumps, shown in the foreground, to enable the introduction of multiple reagents from an auto-sampler to be loaded onto the micro-reactor. Because of the low diffussional distances obtained in this sys-

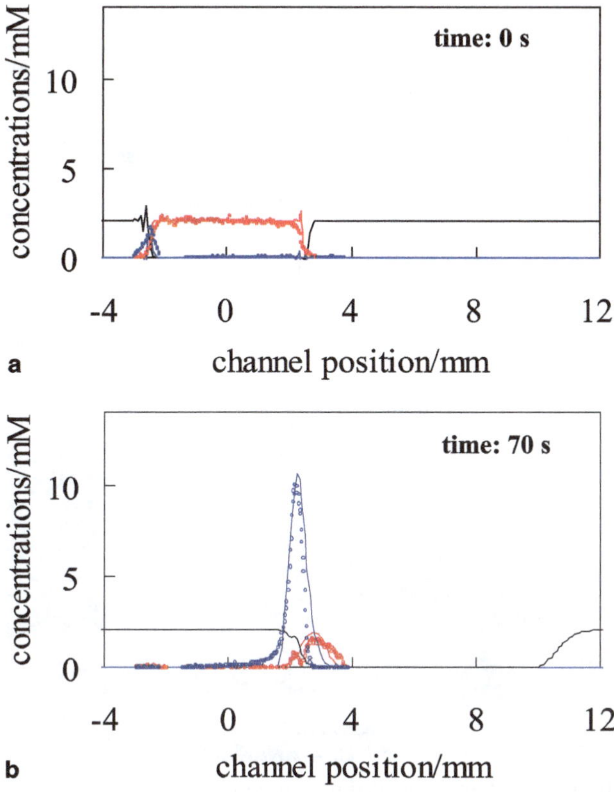

Fig. 5. Concentration profiles of three species involved in a reaction of PADA (pyridine-2-azo-*p*-dimethylaniline) with nickel nitrite to form a complex within a micro-channel. *Solid black lines*, reactant Ni^{2+} concentration; *red points and solid lines*, reactant PADA measured and calculated concentrations; *blue points and solid lines*, product complex measured and calculated concentrations

tem, compared to the use of reagent slugs, this approach is amenable to the individual introduction of a large number of reagents onto the micro-reactor. Hence the rapid loading of sample introduction loops can be performed without compromising either the reaction or analytical flow rates, offering time efficiency whilst ensuring that concentration of the

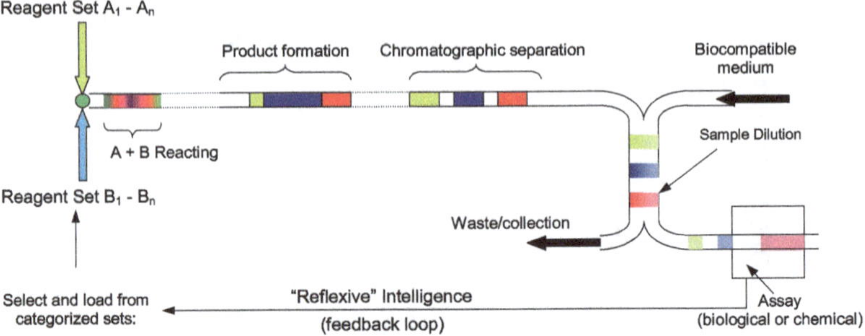

Fig. 6. Conceptual integration of a micro-reactor with a bioassay system

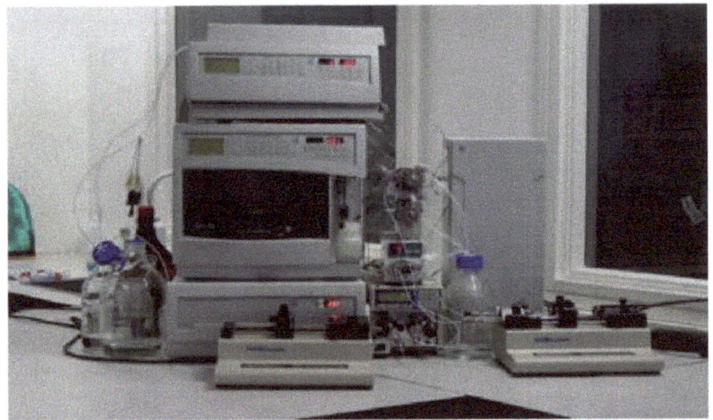

Fig. 7. Illustration of an automated reagent introduction, reaction and analysis set-up based on the incorporation of a micro-fluidic device into a conventional HPLC system

sample is not reduced by excessive diffusion into the carrier solvent. In addition, high flow rates can be maintained for the chromatographic separation, which would be difficult to achieve if the exit stream of the micro-reactor was directly coupled to the analytical column. The ability to independently optimise the flow in all three sections of the process is also important with respect to reagent integrity, reaction efficiency and chromatographic separation. Clearly, the system described has additional applications beyond that of a quality control technique for chemical synthesis; for example, it would be relatively simple to incorporate biological processing and/or couple the reactor set-up to other analytical instrumentation.

6 Summary

It has been demonstrated that miniaturised systems based on microfluidic and Lab-on-a-Chip technology offer many advantages over conventional macro-scale systems, particularly with respect to achieving controllable, information-rich, high-throughput, environmentally friendly and automated processes capable of generating large quantities of product (through parallel scaling) with a high degree of chemical selectivity. These advantages can be attributed to the dramatic reduction in scale leading to unique operating conditions such as the spatial and temporal reagent control obtained under a nonturbulent, diffusive mixing regime and a high surface-to-volume ratio. There is no doubt that micro-fluidic and Lab-on-a-Chip technology can be used as a platform for a wide range of applications such as chemical and biological analysis, chemical synthesis, materials chemistry and biotechnology.

References

Andersson H, van den Berg A (2004) Microfabrication and microfluidics for tissue engineering: state of the art and future opportunities. Lab Chip 4:98–103

Fick A (1855) Uber diffusion. Ann Phys Chem 94:59

Fletcher PDI, Haswell SJ, Pombo-Villar E, Warrington BH, Watts P, Wong SYF, Zhang X (2002a) Micro reactors: principles and applications in organic synthesis. Tetrahedron 58:4735

Fletcher PDI, Haswell SJ, Zhang X (2002b) Electrokinetic control of a chemical reaction in a lab-on-a-chip micro-reactor: measurement and quantitative modeling. Lab Chip 2:102

Fletcher PDI, Haswell SJ, Watts P, Zhang X (2004) Dekker encyclopedia of nanoscience and nanotechnology 2:1547

Fritz JL, Owen MJ (1995) Hydrophobic recovery of plasma-treated polydimethylsiloxane. J Adhesives 54:33

Gillmor SD, Larson BJ, Braun JM, Mason CE, Cruz-Barba LE, Denes F, Lagally MG (2002) Low-contact-angle polydimethyl siloxane (PDMS) membranes for fabricating micro-bioarrays. Proceedings of the 2nd Annual IEEE-EMBS Special Topic Conference on Microtechnologies in Medicine and Biology, Madison WI, USA, p 51

Haber C (2006) Microfluidics in commercial applications; an industry perspectiva. Lab Chip 6:1118–1821

Handique K, Burke DT, Mastrangelo CH, Burns MA (2000) Nanoliter liquid metering in microchannels using hydrophobic patterns. Anal Chem 72:4100–4109

Hau WLW, Trau DW, Sucher NJ, Wong M, Zohar Y (2003) Surface-chemistry technology for microfluidics. J Micromech Microeng 13:272

He P, Haswell SJ, Fletcher PDI (2004) Lab Chip 4:8

Hu SW, Ren X, Bachman M, Sims CE, Li GP, Allbritton NL (2002) Surface modification of poly(dimethylsiloxane) microfluidic devices by ultraviolet polymer grafting. Anal Chem 74:4117

Hunter RJ (1981) Zeta potential in colloid science. Academic Press, London

Jensen KF (2001) Microreaction engineering: is small better? Chem Eng Sci 56:293

Lagally E, Mathies RA (2004) Integrated genetic analysis: microsystems. J Phy D Appl Phys 37:R245

Laurell T, Nilsson J, Jensen K, Harrison DJ, Kutter JP (eds) (2004) 8th International Conference on Miniaturized Systems for Chemistry and Life Sciences (MicroTAS 2004). Malmö, Sweden, September 26–30

Manz A, Becker H (eds) (1998) Microsystem technology in chemistry and life sciences. Springer, Berlin Heidelberg New York

Metaxas AC, Meredith RJ (1983) Industrial microwave heating. Peter Peregrinus, London

Mills R (1973) Self-diffusion in normal and heavy water in the range 1–45°. J Phys Chem 77:685

Munson MS, Hasenbank MS, Fu E, Yager P (2004) Suppression of non-specific adsorption using sheath flow. Lab Chip 4:438–445

Munson MS, Hawkins KR, Hasenbank MS, Yager P (2005) Diffusion-based analysis in a sheath flow microchannel: the sheath flow T-sensor. Lab Chip 5:856–862

Nikbin N, Watts P (2004) Solid-supported continuous flow synthesis in microreactors using electroosmotic flow. Org Process Res Dev 8:942

Overbeek JTG (1952) Electro chemistry of double layer. In: Kruyt HR (ed) Colloid science, Vol. 1. Elsevier, Amsterdam

Peterson DS, Rohr T, Svec FK, Frechet JMJ (2003) Dual-function microanalytical device by in situ photolithographic grafting of porous polymer monolith: integrating solid-phase extraction and enzymatic digestion for peptide mass mapping. Anal Chem 75:5328

Rice CL, Whitehead R (1965) Electrokinetic flow in a narrow cylindrical capillary. J Phys Chem 69:4017

Schneider TW, Schessler HM, Shaffer KM, Dumm JM, Younce LA (2001) Surface patterning and adhesion of neuroblastoma X glioma (NG108-15) cells. Biomed Microdev 3 4:315

Svec F (2004) Porous monoliths: emerging stationary phases for HPLC and related methods. LC GC Europe 18:17

Takagi M, Maki T, Miyahara M, Mae K (2004) Production of titania nanoparticles by using a new microreactor assembled with same axle dual pipe. Chem Eng J 101:269

Takayama S, McDonald JC, Ostuni E, Liang MN, Kenis PJA, Ismagilov RF, Whitesides GM (1999) Patterning cells and their environments using multiple laminar fluid flows in capillary networks. Proc Natl Acad Sci USA 96:5545–5548

Whitesides GM (2006) The origins and the future of microfluidics. Nature 442:368–373

Yang YN, Li C Kameoka J, Lee KH, Craighead HG (2005) A polymeric microchip with integrated tips and in situ polymerized monolith for electrospray mass spectrometry. Lab Chip 5:869

Zhang X, Hayward DO (2006) Applications of microwave dielectric heating in environment-related heterogeneous gas-phase catalytic systems. Inorg Chim Acta 359:3421

Zhang X, Wiles C, Painter SL, Watts P, Haswell SJ (2006) Microreactors as tools for chemical research. Chim Oggi-Chem Today 24:43

Ernst Schering Foundation Symposium Proceedings, Vol. 3, pp. 39–55
DOI 10.1007/2789_2007_027

Published Online: 23 May 2007

Microreactors as New Tools for Drug Discovery and Development

S.Y.F. Wong-Hawkes, J.C. Matteo, B.H. Warrington(✉), J.D. White

BB Consultants Ltd, 45 The Drive, SG14 3DE Hertford, UK
email: *brianwarrington@btconnect.com*

Abstract. In common with other producers of fine chemicals, the pharmaceutical industry can deploy flow microreactors to provide a more flexible production regime than is achievable with large-scale batch reactors. With monitoring of output flow, microreactors can be adjusted exquisitely to give enhanced and inherently safer protocols. Bulk production can be achieved through long run times or parallel reactors. However, microreactors can also be used advantageously in pharmaceutical R&D in other ways. This paper describes their use in the creation of a tool for rapid discovery and optimisation of leads to new drugs. Here microreactors are used to create an integrated micro-scaled chemical synthesis and bioassay system able to conduct fast-cycling iterative searches of diverse chemical space as an enhanced method for identifying and optimising novel lead chemotypes. The use of a microreactor to provide point-of-use access to PET ligands is also described. This significant down-scaling and acceleration of the synthesis of potent but short-lived biomarkers of disease could provide a means of significantly shortening the time and resources required to achieve a drug approval.

1 Introduction

There are many reasons why chemical synthesis is advantageously performed in flow mode using a micro-contactor (or microreactor) rather than in a round-bottomed flask, well or vessel. In fact, if it was not for a long history of batch mode chemistry and the convenience of a hand-sized flask, the case would need to be made for employing batch methods.

The flow microreactor has several advantages for those wanting to make substantial amounts of a product on a regular basis, particularly in terms of process enhancement and safety. Because of the small cross-sectional dimensions of the of the reaction zone, mixing occurs within a relatively small number of molecular collisions, thus ensuring much faster, more efficient and more reproducible mass and heat transfer than is achievable in a batch reactor. The effects of erratic mixing and thermal gradients within the reaction bulk are thus largely avoided. The micro-environment can also bestow other beneficial effects on reaction outcome due to the influence of flow on the equilibrium constant and the catalytic effect of interfaces due to the huge increase in surface area-to-volume ratio in these small systems. In continuous flow mode, it also allows tighter control of product quality and safety by providing the ability to monitor output identity and quality and to apply this feedback in real time to optimise the operating conditions. Output monitoring can be used to shut off flow if the product is not detected. As only a small amount of chemistry is in play at any time, this leads to an inherently safer process and avoids the dangers that exist in the batch method, where accumulation of added reagents can release substantial amounts of energy over a short time when the reaction eventually initiates. The production bulk product is achieved by running a single microreactor for a sufficiently long time or, if time is critical, several identical reactors in parallel, although each must be monitored.

Overall, the microreactor provides greater safety for individuals and equipment and reduces the likelihood of loss of process and the consequent disruption and even loss of sales that can follow. In common with other fine chemical manufacturers, most pharmaceutical companies have programs to capture the benefits of flow microreactors as adjuncts to or even replacements for their current batch methods for scaling up production of candidate molecules to satisfy clinical and manufacturing needs. This paper attempts to demonstrate that microreactors can be deployed more widely in pharmaceutical R&D than as a tool for enhanced production and that they have the potential to underpin significant paradigm shifts in both early- and late-phase R&D.

2 The Challenges Facing the Pharmaceutical R&D Process

It can take from 10–15 years from target discovery to FDA approval to produce a new drug. (US Food and Drugs Administration 2004). This drug will be a survivor amongst many failed endeavours; target attrition is over 90%. It is therefore not surprising that the cost of each R&D cost per approval is about US$1.2 billion, with much of that set to defray the cost of attrition. Furthermore, R&D productivity in the larger pharmaceutical companies has actually fallen in the last decade despite increased investment. During the period 1996 to 2004, the number of FDA approvals fell by more than half whilst R&D spending more than doubled. As a result, price-based rationing of pharmaceutical products through governmental or reimbursement agencies is now becoming commonplace, even for novel products with unprecedented benefit. The revenue of pharmaceutical companies is also under threat. The industry faces shorter and shorter periods of market exclusivity for first-in-class drugs. During the 1980s, a company could expect 5 years of market exclusivity before a rival product appeared on the market. By the 1990s, that period had been reduced to 1 or 2 years, and today a competitive product can appear within months of the primary launch (DiMasi and Paquette 2004). The future holds poor prospects, since within the next

5 years, 115 drugs which currently generate US $100 billion in annual revenue will come off patent. This represents a substantial portion of a US $430 billion global pharmaceutical industry. Although the life cycle management and product-line extension techniques currently employed by the pharmaceutical and biotechnology companies will help preserve a small portion of this revenue, the risk is that generic competition will take the lion's share.

Although for more than 20 years individual large pharmaceutical companies have attempted to counter the issue of decreasing output in a competitive manner, their strategic solutions have been remarkably similar. Usually what has been proposed is a move from local endeavours and knowledge directed to incremental improvements in a particular therapeutic area to a search for block-buster products through a centralised corporate wisdom based on creating and accumulating knowledge of targets based through bioinformatic and cheminformatic approaches. It was considered that only this approach could deliver the novelty demanded by regulatory authorities and the revenue that pharmaceutical companies needed to survive. Important features in this approach were the establishment of central resources for creating chemical diversity, conducting high-throughput screening and for the curation and mining of informatics databases. As the benchmark figures show, this has not been a great success. As well as delivering reduced productivity, the industry has undergone substantial consolidation through mergers and acquisitions.

Long cycle times and high costs mainly reflect the 90% attrition; therefore this is the key problem to address. The proportion of targets lost through failure to reach the screening stage is actually quite small and failure as early as this has a relatively low cash cost. Most attrition is due to late-stage failure of an advanced chemical lead or a development candidate to exceed the efficacy and toxicity criteria required for progression. Failure at these later stages is very costly and disruptive to plans and effort deployment. While one conclusion could be that for some targets perhaps there is no chemical solution, this is unlikely as frequently where one company fails another succeeds. It is more likely that the cause of attrition is a deficiency in the way the chemical lead or candidate is identified.

3 Use of Microreactors to Enable Iterative Lead Optimisation

It is likely that the cause of attrition is rooted in a poor choice of the lead series that are subjected to the optimisation process. These are selected by assessment of the hits yielded by high-throughput screening for their chemical tractability and by a limited examination of their potential for side effects. Although many screening hits are identified, chemistry resources will rarely stretch beyond advancing one or two as leads for optimisation. Although the intent of high-throughput chemistry has always been to improve the chemical diversity available for lead selection from screens or to identify diverse structures during lead optimisation, the practice has usually been to target large numbers of compounds by using combinatorial and parallel synthesis. These methods, of course, derive their enhanced numeric productivity from the use of a common method to create many related compounds. It is therefore doubtful whether compound collections used in screening to identify lead series show an increase in diversity commensurate with their rapid growth in recent years. In fact, the historic segment of the collection comprised of hand-made compounds is usually the richest source of leads. Thus, although the collection sizes approach the low millions, they contain many large clusters of similar compounds and it is unlikely that more than a few thousand compound classes are represented, whilst more than 10^{40} chemotypes that might reasonably described as druggable (Lipinski et al. 1997) could be considered to exist.

Not only does the limited diversity of the compound database curtail the choice of lead, the process for assessing screening hits to see if they define a tractable lead is also limited by resources, particularly in those specialist assays that would predict the ability of the chosen lead class to successfully yield an approval in a particular therapeutic field. The process is therefore driven mainly by consideration of activity against the target. Therefore chemotypes with adequate but less than maximal potency for the target but freedom from damaging side activity remain undetected in this process.

Thus, for the average chemist the typical annual output of 10–50 carefully hand-crafted compounds of the 1970s and 1980s had become over 1,000 by the late 1990s and for those specialising in high-throughput chemistry methodology, tens of thousands per individual would not be unusual. Less obvious is the price paid for the change in terms of strategy. The low throughput of the earlier campaigns did not overload assay capacity and it was possible to perform many secondary assessments in the timeframe similar to that required to produce a compound. These assays would sometimes include some powerfully predictive assessments of efficacy and toxicity and thus contribute to the direction of the chemistry and thereby increase the probability that a productive course was being followed. The chemist was therefore able to redesign each compound in response to a range of biological information received. Therefore, right from the start the chemist was delivering compounds targeted to the needs of a specific therapeutic target and each revision of the lead structure posed a specific and relevant new hypothesis. In fact diversity was achieved by default as the outcome of attempts to probe receptor space. In contrast, the early stages of the modern lead discovery process do not allow for such an iterative approach because of batch delivery of compounds. Redesign is now at the rate of one library at a time, rather than one compound, because the numeric efficiency of the high-throughput method is present only if the production run is continuous. Therefore any opportunity to guide biological optimisation by applying feedback from biological assays is denied.

Overall a strategic change has been made from serial iterative chemistry being guided by essentially parallel biological assays to the reverse: parallel chemistry serving assays which deliver their results serially over a long period of time. The change was driven by a perception that the earlier methodology was inefficient because of its slowness. However, the superiority of the iterative closed loop search algorithm over the open loop declarative method is nicely illustrated by comparing their potential performance in an imaginary situation (Kramer et al. 1998). Approximately 1,750 diamines can be purchased from commercial sources, as can 26,700 carboxy synthons, which could be used to synthesise about 10^{12} diamides, some of which will be known drugs in clinical use. To find any one drug, the combinatorial approach would require making them all. This would require about 10^{12} 40-well plates as for synthesis and a similar number of assay plates to provide a complete data set of results. At 1 mg per well, roughly 100 million kg of reagents would be required. In contrast, if the biological data were sharp enough to choose between two alternative structures for the next diamide to be synthesised, then only 40 compounds need to be synthesised in an iterative search. The difference in performance may go some way in explaining why the early discovery campaigns of the 1960s and 1970s would optimise a lead to a commercial entity via a series of approximately 1,000 molecules, whilst that level of output is routinely achieved within a day in the modern environment without yielding a drug.

Microreactor-based technology may challenge the current paradigm because the low-cost, high-speed compound handling capacity that can be achieved in chemistry and assays may permit the return of the iterative method. For the pharmaceutical companies, an early-stage, low-operating-cost, drug discovery platform that can quickly identify leads with an enhanced immunity to attrition and a greater likelihood from a larger pool of hit options stands a better chance of providing best-in-class therapy than the current methodology.

4 Micro-Scaled Lead Generation and Optimisation Systems

Fast serial iterative chemistry coupled with similarly fast parallel assays can be enabled by deploying flow micro-reactors in both chemistry generation and the assays. The parallel assays would be used to assess potency at the target protein and effects in assays that would define cross-reactivity, toxicity and efficacy. Such systems have cycle times normally measured in minutes and with feedback of assay data to determine reagent selection and thus compound design could plot a course for the chemistry that acknowledges several concurrent SARs and therefore provide a way of integrating several steps of the current serial biological assessment regime. Not only would the overall cycle time be reduced but the full deployment of secondary assays even back into the hit to lead process would give greater assurance that the best lead series (not necessarily the most potent in high-throughput screening) had been selected and the risk of attrition thereby reduced.

However, adoption would require the acceptance of a different strategy for lead discovery and possibly also lead optimisation. The key changes required would be that beyond the high-throughput screening stage, each biological target would be treated as a separate entity and make no further call on centralised processes or resources such as registration, compound management or possibly even open-loop high-throughput synthesis. In this way material transfer and waiting times would be ruthlessly eliminated from the iterative cycle. To be effective, additional changes would be required. For example, an understanding that no compound would be isolated for local storage and the amount made would be only that required for the assays, which for a modern sensitive assay can be in the region of picograms. This would greatly reduce all local process times and relieve issues associated with chemical intermediate supply chains.

To test the feasibility of this proposition, a chemistry-generating system and product-separation system were joined together as the concatenated components of an effectively continuous microfluidic system, thus virtually eliminating the possibility of loss of time or materials in handling or transfer. Most of the chemistry-generating system resided in a fused glass chip which was etched to form the microreactor junctions and a serpentine flow channel to provide a longer residence time for the reactive species. As the system was planned to function almost at a "no mass–information only" level, very small channel cross-sections were used (e.g. 50 × 15 μm). As these were expected to deliver very short reaction times (possibly down to seconds), the system as a whole was coordinated electronically from a computer through a LabView program. The system is shown schematically in Fig. 1 and its reality is depicted in Fig. 2 and with the exception of the custom-designed chips is mainly an assembly of commercially available equipment. In normal use, the chemistry reagents are injected into the system and are propelled by the pumping system. The pump needs to be highly reproducible in its actions as chip volumes are very small compared to pump capacity. When the reagents reach the chip they mix and react. The reaction time depends on the combination of flow rate and channel size and length. Usually, a make-up flow is added just before the outlet of the chip, which can either quench the reaction and/or move the slug faster through the rest of the system. On elution, the product slug is detected by an UV-Vis detector. The detection of the slug triggers an injection of part of the slug into the LC system where reaction components are time resolved. The flow of products is monitored by a UV detector and prod-

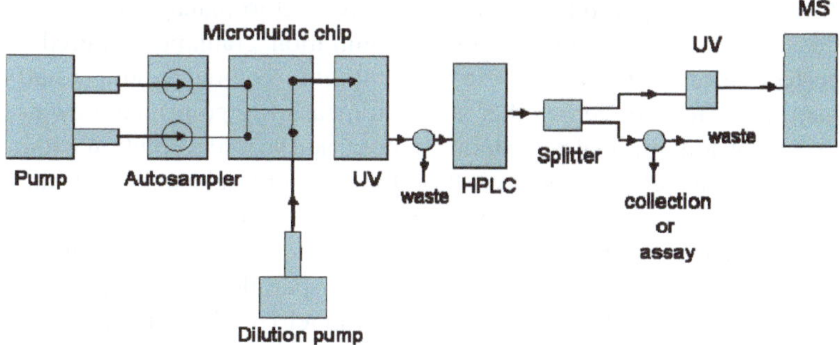

Fig. 1. Schematic diagram of lead discovery and optimisation system

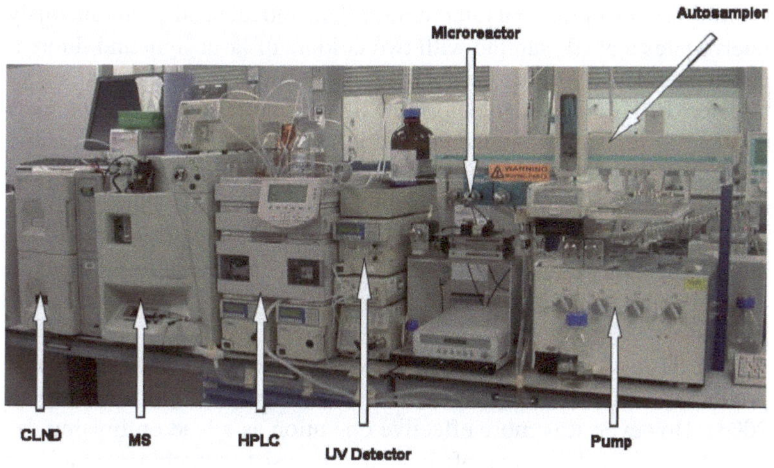

Fig. 2. Lead discovery and optimisation system

ucts are identified by mass spectroscopy. The appearance of a product with appropriate preset properties triggers an entry into the automated flow assay system.

A flow assay system (Fig. 3) enables ligand binding and functional activity measurements in glass channels with dimensions approximately 20 × 15 μm. It has a temperature-controlled stage to hold the chip.

A four-channel pump and valve system is used to manage fluid flows through the chip to generate the concentration gradients required to perform the titration to find the concentration required to give a half-maximal effect (or some other standard measuring point). As flow assays show longitudinal and lateral temporal and spatial resolution within the flow channel, a suitable position for reading a relevant event must be found and locked. Therefore, the chip is mounted on a closed loop three-dimensional motion control system which can move the optical spot relative to the channel over a 20-mm span down to an accuracy of less than 2 μm. In this way, a flow channel length over 1 m length formatted as a serpentine within the area of the chip can be monitored. The system uses a patented focussing system (White 2003) to locate the channels and the focal depth, which obviates the need for an optical microscope or manual intervention. The optical head simultaneously interrogates a 30-fL sample with two colours of laser light and fluorescence intensity, fluorescence polarisation or fluorescent resonant energy transfer (FRET) is used to determine ligand binding. Sensitivity ranges down to as few as 400 molecules in the sample volume at any one time and under some circumstances a single fluorescent molecule can be detected. With this degree of scale-down and the fact that valuable target protein is only released when required and in the quantity actually required, the assay device is extremely frugal with respect to protein and uses only about one-thousandth the amount required for microtitre plate well assay.

This basic chemistry generator system has proved itself to be adept at producing a diverse range of chemistry (e.g. Garcia-Egido et al. 2002, 2003). However, it is more effective operation as a lead optimising device if certain tasks can be off-loaded to two sister machines which act in concert with the main device. These are similar in their configuration to the main machine but are adapted to achieve their specific roles. Thus a reaction validation and optimisation platform (VOP; Fig. 4) takes this role offline.

As well as ensuring loss-free handling of microgram down to picogram amounts of materials from reagent to assay and ensuring very short reaction times (usually in the region of seconds to low minutes) due to short diffusion paths, stable longitudinal and lateral (1.5-dimensional) spatial and temporal separation of materials in a low Reynolds

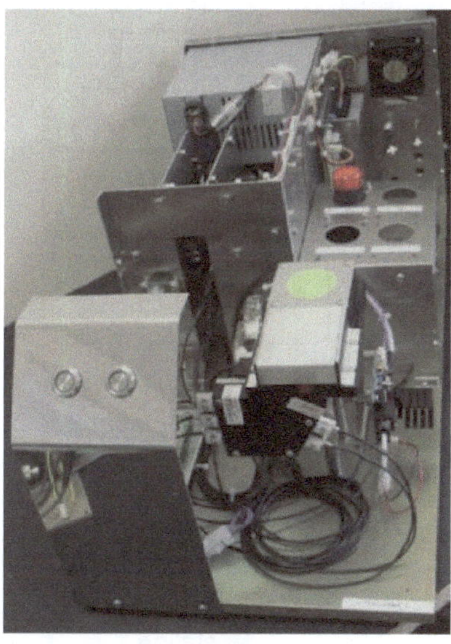

Fig. 3. Flow assay device (courtesy of Genapta Ltd.)

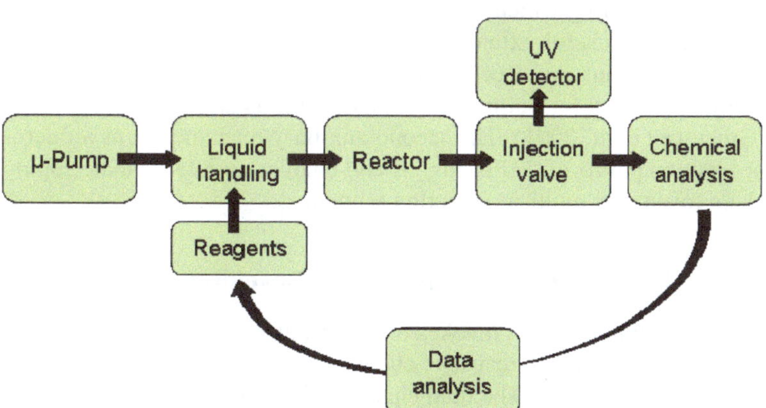

Fig. 4. Validation and optimisation platform configuration

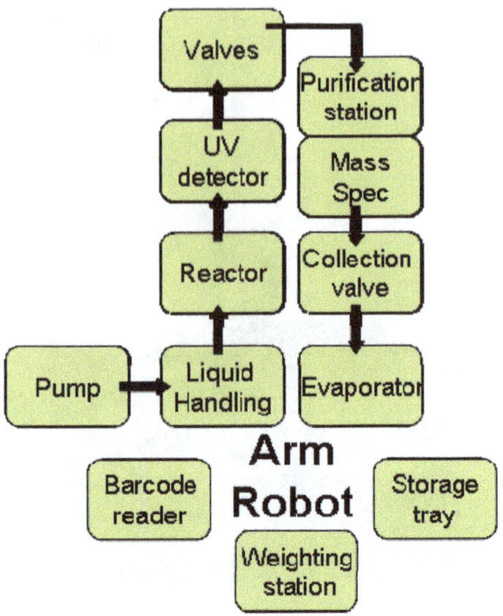

Fig. 5. Micro-production platform configuration

number tube provides the chemist with an ability to partially control reagent contacting which is not available in a (zero-dimensional) flask or well. To a useful extent, changing the conditions within the VOP microreactor has an effect on outcome and this is used to good effect in searching for optimal reaction conditions and changes in output (Greenway et al. 2001). The second auxiliary machine forms a micro-production platform (MPP; Fig. 5) and enables an intermediate supply chain to support multistep reaction sequences.

5 Microreactors to Enhance the Development Process

The development phase invokes many instances where reactions must be optimised quickly because of changes in the route to or specification of the drug substance. The final challenge is to converge upon the most appropriate and environmentally acceptable methods for manufacture. In identifying optimal synthesis conditions, real-time tuning of a microflow reactor conditions while monitoring outflow is faster and more effective than carrying out a huge number of well- or flask-based trials. The proportion of precursor used in establishing the best conditions relative to that which will be harvested as product will also be lower. As with the drug discovery application, real-time optimisation in flow is essentially an iterative process, as successive changes to reaction conditions are based on using the results as they accumulate. In contrast, well-based design of experiment methods bear some similarity to the parallel array strategies, as the set of experimental conditions deemed important enough to explore is prejudged prior to the experiment. The real bonus with tuneable flow technology is that new mechanisms and parameters that contribute to the identity of products and their yield and purity can be learned.

Benefits can also accrue simply from the way the reaction is being conducted. For example, provided the Reynolds number is low, the conditions in a small channel are very well-behaved, stable and reproducible. In contrast, the stochastic nature of the start-up and quenching of a large batch reactor will always introduce a random factor between one batch and the next. Flow micro-reactors can also inherently provide a tighter ordering of events in the reactor. For example, for a nonturbulent flow stream of reactants passing through a reactor hot zone in a micro-channel, by and large the first reagent molecules to enter form the first product molecules to exit the hot zone, and all reagents and products experience the same reaction conditions for the same length of time. In contrast, there is no guarantee that the first reagent molecules to be added to a large heated and stirred batch reactor go to form the first product molecules to be quenched or worked up. This variation in the length of exposure to reaction conditions can have a significant effect on the yield and quality of the product, even to the extent of providing the only viable method.

6 Microreactor Synthesis of PET Biomarkers in Development

As well as enhancing the production of drug substance, micro-reactors can be used to assist the development process in a different way: that of enabling the powerful technique of positron emission tomography (PET) to be readily deployed as a point-of-use tool in clinical trials and the general biology lab. Direct PET imaging of the distribution, metabolism and elimination of novel drugs, particularly in human subjects, can reduce the development cycle time by many months. In fact, when performed close to a local cyclotron source, the micro-reactor may be able to supply samples of such high specific activity that the near zero physical dose is required. Under these circumstances, a much earlier approval for dosing to humans could be possible.

PET biomarkers are a critical tool for drug discovery and the clinical detection of cancer, heart disease, Alzheimer's disease and other disorders. The gold standard PET probe, [^{18}F]fluorodeoxyglucose, is used to monitor glucose metabolism in the body. Other more specific biomarkers, e.g. [^{18}F]fluoromisonidazole, a marker for hypoxia, and [^{18}F]fluorothymidine, a marker of proliferation, and others have been developed, but are extremely difficult and slow to bring to market. Using conventional chemistry, these specialised biomarkers are difficult to synthesise, often have poor yields, and require expensive precursors. A flow-based microfluidic chemistry system has been developed to rapidly produce and deliver PET-labelled biomarkers for in vivo imaging. A modular flow-based micro-reactor assembly capable of four reaction steps has been developed to provide a compact multistage reactor device (Fig. 6). Only glass and PEEK wetted parts were used to ensure solvent tolerance and the ability to add heat to accelerate reactions.

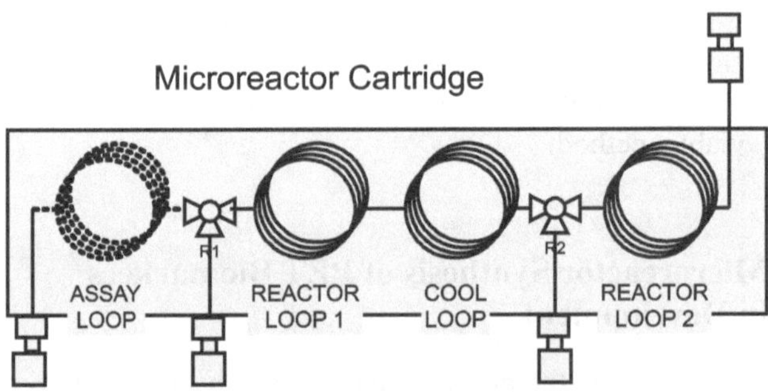

Fig. 6. Four-stage flow-based reactor for PET biomarker synthesis

The resulting device has demonstrated both FDG and FLT labelling at yields of 98% and 90%, respectively, in 100 s compared to typical macro-scale labelling of 65% in 45 min for FDG and 30% in 90 min for FLT. The use of acetonitrile, DMSO and HCl have shown no degrading effect on the system. Extremely efficient labelling illustrates the effectiveness of flow-based micro-reactors for PET biomarker synthesis. Multiple biomarkers can be produced in 1–2 min, while using only micro-litres of precursor and can revolutionise the production of radiotracers. Small reaction volumes, improved yields, and the ability to synthesise small quantities of a variety of new compounds will allow preclinical and clinical evaluation of new PET agents with potential for clinical utilisation.

A biomarker generator system using a micro-accelerator with micro-fluidic chemistry is being developed to rapidly produce and deliver PET-labelled biomarkers on demand and at the point of use for in vivo imaging. The biomarker generator for positron emission tomography (PET) produces single-dose quantities of [^{18}F]fluoride ion (15 mCi) and then converts that into a PET biomarker dose approximately every 15 min. The combination of a compact low-cost accelerator with fast and efficient micro-fluidic chemistry located near the imaging site eliminates a major portion of the cost and inflexibility of the present distribution networks for FDG and other biomarkers for PET. Doses would be produced on demand in single-dose batches that enable the synthe-

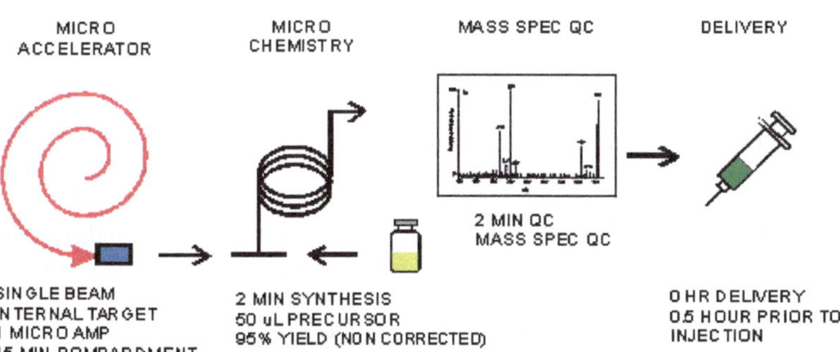

Fig. 7. Proposed biomarker generator for producing [F-18]-labelled PET biomarkers produces single-dose batches on demand

sis of multiple doses of different PET biomarkers with no impact on efficiency. Figure 7 illustrates the proposed PET biomarker generator system for producing [F-18]-labelled PET biomarkers produces single-dose batches on demand. The accelerator produces approximately 0.5 mCi/min and total chemistry and QC processing takes 5 min.

7 The Prospect of Nano Reactors

Obviously a truly nano-scaled flow reactor is impossible because it will be incapable of bulk flow. However, the concept is a useful one as a flow micro-reactor, probably with a submicron bore, could be used to deliver to assays implemented at the nano scale where factors associated with a protein single molecule can be discerned. In general, single-molecule assays provide more information than a solution-based assay and the origin of the information is more certain as the molecular source is fully characterised. The information content is also enhanced because even when a bulk solution assay contains only one type of biological entity, discerning the signal of characteristic molecular events is not possible because of the asynchronous output from the many molecules present. These assays, because of their directness and high information content, will provide biomarkers to those currently employed and thus contribute to the reduction of cycle time and attrition in pharmaceutical R&D.

References

DiMasi JA, Paquette C (2004) The economics of follow-on drug research and development. Pharmacoeconomics 22(2):1–14

Garcia-Egido E, Wong SYF, Warrington BH (2002) A Hantzsch synthesis of 2-aminothiazoles performed in a heated microreactor system. Lab Chip 2:31–33

Garcia-Egido E, Spikmans V, Wong SYF, Warrington BH (2003) Synthesis and analysis of combinatorial libraries performed in an automated micro reactor system. Lab Chip 3:73–76

Greenway GM, Haswell SJ, Styring P, Morgan DO, Skelton V, Warrington BH, Wong SYF (2001) The generation of concentration gradients using electroosmotic flow in micro reactors allowing stereoselective synthesis. Analyst 126:11–13

Lipinski CA, Lombardo F, Dominy BW, Feeney PJ (1997) Experimental and computational approaches to estimate solubility and permeability in drug discovery and development settings. Adv Drug Del Rev 23:3–25

US Food and Drug Administration (2007) Challenge and opportunity on the critical path to new medical products. http://www.fda.gov/oc/initiatives/criticalpath/whitepaper/html, Cited 6 February 2007

White JD (2003) A dual wavelength optical fluorescence analyser, WO/2003/048744

Ernst Schering Foundation Symposium Proceedings, Vol. 3, pp. 57–76
DOI 10.1007/2789_2007_028

Published Online: 24 May 2007

Microchemical Systems for Discovery and Development

K.F. Jensen(✉)

Department of Chemical Engineering, Massachusetts Institute of Technology,
77 Massachusetts Avenue, 02139 Cambridge, USA
email: *kfjensen@mit.edu*

Abstract. Applications of silicon-based microreactors are summarized starting with systems for single-phase organic transformations and progressing through multiphase catalytic systems to microsystems for multistep chemical synthesis. The latter systems involve extraction and gas–liquid separation processes designed to take advantage of the dominance of surface tension effects in microfluidic devices. Integration of physical sensors (e.g., for pressure, temperature, and flow) and measurements of chemical species further enhances the utility of microreactors by enabling chemical kinetic studies and optimization of optimal operating conditions. A brief description of synthesis and handling of solid particulates is included, with particular emphasis on multistep processing of colloidal nanoparticles. Finally, scale-up issues and challenges to the adoption of microreaction technology are discussed.

1 Introduction

Microreactors are continuous-flow chemical synthesis devices with submillimeter width fluid channels. In their most basic form, they consist of inlet, mixing, reaction, and outlet sections (Fig. 1). Advanced microreactor devices also include sections for concentrating and capturing reagents (Lee et al. 2005) as well as physical (de Mas et al. 2005) and chemical sensors (Jackman et al. 2001; Floyd et al. 2005; Löbbecke et al. 2005). During the past decade, microreactors have evolved from early prototypes applied to relatively simple chemistry examples to promising laboratory techniques used in multistep synthesis (Ehrfeld 1997; Ehrfeld et al. 2000; Fletcher et al. 2002; Jähnisch et al. 2004; Pennemann et al. 2004; Lee et al. 2005; Baxendale et al. 2006). The choice of material for microfluidic devices depends on chemical compatibility, temperature, and pressure, as well as ease of fabrication and integration. Microstructures fabricated in metals are compatible with organic solvents and can be operated at elevated temperatures and pressures, but they are not suitable for reactions involving strong acids and bases (Pennemann et al. 2004). Ceramic-based devices are chemically inert and stable at high temperatures, but microfabrication is complicated by matching shrinkage (typically ~10%) and thermal expansion during firing (Knitter et al. 2001). Glass is the material of choice for many chemical reactions, but creating deep three-dimensional channel structures in glass is difficult (Fletcher et al. 2002; Kikutani et al. 2002). Polymer-based microfluidics, specifically rapid prototyping in poly(dimethylsiloxane) (PDMS) (soft lithography) (Xia and Whitesides 1998), have cost advantages, but the materials dissolve in or are swelled by common organic solvents (Lee et al. 2003).

Examples of using metal, polymer, and glass microreactors appear in other chapters of this volume. The present chapter focuses on microreactors created in silicon, a material that has high mechanical strength,

▶

Fig. 1. General microreactor for organic transformations as the round-bottom flask equivalent. The microreactor incorporates (i) a laminar flow mixing section of separate streams, exemplified for glycosylation, (ii) reaction channel, inlet for quench, and an outlet (Ratner et al. 2005; Murphy 2006)

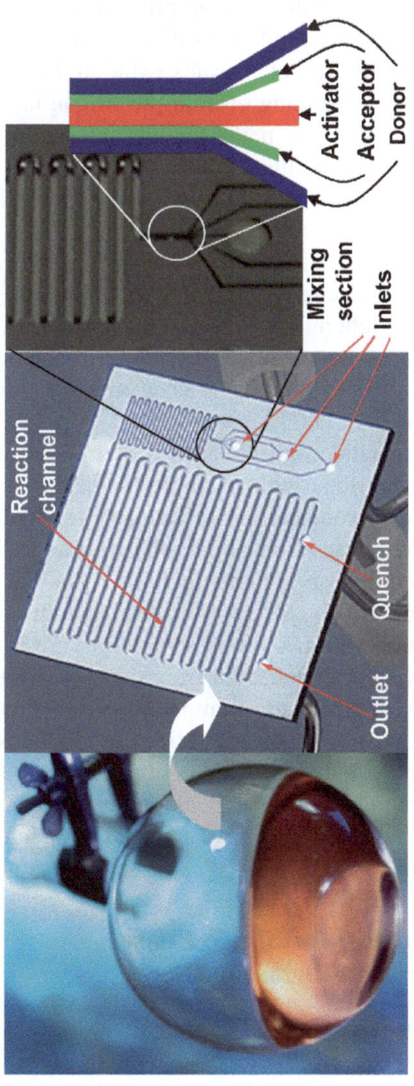
Reaction channel
Quench
Outlet
Mixing section
Inlets
Activator
Acceptor
Donor

excellent temperature characteristics, and good chemical compatibility. Well-established wet and dry etching procedures enable fabrication of microchannels with controlled sidewall shape and channel dimensions from micrometer to millimeter scales (Madou 2002). Silicon–silicon wafer bonding allows bonding of multiple silicon layers to form complex three-dimensional flow paths feeding into and exiting from specially designed reaction volumes (Wada et al. 2006). Anodic bonding of Pyrex and silicon provides a strong seal of the microfluidic channels and visual access to reaction flow channels. Moreover, oxidation of the silicon forms a glass layer on the surface so that oxidized silicon microreactors become functionally equivalent to glass equipment used in bench-scale chemistry experiments. For cases where silicon or glass lacks the necessary chemical resistance, it is possible to incorporate alternative protective coatings such as nickel for direct fluorination reactions (de Mas et al. 2003).

Applications of silicon-glass-based microreactors are summarized in the following sections with examples selected to illustrate potential advantages of microreactors in chemical research and development. The discussion progresses from single-phase organic transformations to more complex multiphase systems including gas–liquid reactions and synthesis of nanoparticles. Microfabricated separation systems, specifically for extraction and gas–liquid separation, are included to enable microreactor-based continuous multistep synthesis. Integration of physical sensors (e.g., for pressure, temperature, and flow) and measurements of chemical species is emphasized as a means to enhance the utility of microreactors by allowing chemical kinetic studies and optimization of operating conditions. Scale-up and challenges to the adoption of microreaction technology are touched upon in the final section.

2 Microreactors for Single-Phase Organic Transformations

The microreactor shown in Fig. 1 can be regarded as a continuous-flow reactor equivalent to the common batch reactor flask. The microreactor has a mixer section that provides more precise mixing characteristics than those typically achievable in a stirred vessel. For example, in the case of glycosylation (Ratner et al. 2005) the mixing by diffusion characteristics of laminar flow allows "stacking of the fluids" so that the donor only reaches the activator in presence of the acceptor (cf. Fig. 1), and side reactions are then minimized. The mixing time clearly must be less than the reaction time in order to realize the benefit of the microreactor operation. Consequently, a multitude of mixer designs have been demonstrated for achieving fast mixing (Ottino and Wiggins 2004), including fluid lamination (Ehrfeld et al. 1999), chaotic mixing (Stroock et al. 2002), and mixing by introducing a segmenting fluid (Song et al. 2003; Günther et al. 2005). The average time a fluid volume spends in the reaction channel (the residence time) is equivalent to the reaction time and can be manipulated by changing the volumetric flow or length of the channel. A quench inlet at the end of the reaction makes it possible to confine the reaction to the device, which becomes particularly important when using microreactors to determine the reaction kinetics and to establish optimal reaction conditions. The outlet tubing otherwise adds a significant volume and uncertainty as to whether the reaction is completed in the device or in the interface to the macroscopic fluid-handling environment.

Connecting the microreactor to external fluid reservoirs is a critical engineering challenge in transforming microreactors from demonstration studies to useful laboratory technology (Fredrickson and Fan 2004). Epoxy sealing, often used with polymer devices, is adversely affected by organic solvents. Compression sealing fluid components to a macroscopic holder by elastomer gaskets creates potential leakage problems and adds mechanical stress that can lead to chip fracture. For cryogenic temperatures to moderate temperatures (–80° to ~250°C), robust and compact seals can be achieved by soldering techniques inspired by "flip-chip" bonding methods used in computer chips (Murphy 2006). Combinations of Pyrex, Kovar tubes, and glass brazing produce stable connections for high temperature (<500°C) and pressure applications (<200 bar) (London et al. 2001).

The microreactor in Fig. 1 demonstrated the ability to rapidly obtain comprehensive information about a given chemical transformation, e.g., glycosylation, over a wide range of conditions (Ratner et al. 2005). Experimental efforts and time were considerably reduced and valuable starting materials conserved by combining the microreactor with automated HPLC analysis. With a single preparation of reagents, 44 reactions were completed in less than a day at varying temperatures and reaction times requiring just over 2 mg of glycosylating agent for each reaction. Conventional batch reaction procedures were more than ten times slower, limited to approximately three experiments per day, and required significantly larger quantities of starting material.

3 Gas–Liquid (Catalyzed) Reactions

Continuous microfluidic reaction technology is particularly advantageous for multiphase reactions by typically providing hundred-fold enhancement in heat and mass transfer. The actual extent of increase in interfacial transport depends on the nature of the multiphase flow and the geometry of the microfluidic channel (Günther and Jensen 2006). Including solid particles or microfabricating structures (e.g., pillars) (see Fig. 2) further increases interfacial areas for reactions (Losey et al. 2002; Wada et al. 2006). The higher mass transfer rates in microreactors enhance gas–liquid reactions, such as heterogeneous catalytic hydrogenations (Losey et al. 2001, 2002; Kobayashi et al. 2004), typically limited by transport of gas species into to the liquid reaction medium in macroscopic reactors.

The much larger rates of heat transfer in microreactors allow highly exothermic reactions that are difficult to control in conventional batch systems, such as direct fluorination (Chambers and Spink 1999; Jähnisch et al. 2000; de Mas et al. 2003) and ozonolysis (Wada et al. 2006). Thus, microreactors expand the number of possible reactions and feasi-

▶

Fig. 2. **a** Microreactor for catalyzed gas–liquid reactions (e.g., hydrogenation): *right insert* shows interleaving of gas–liquid inlets to increase interfacial area; *left insert* shows gate structure holding catalyst particles in place in channel (Losey et al. 2001). **b** Ten-channel version of microreactor for catalyzed gas–liquid reactions: *left insert* 50 μm catalyst particles loaded in channel; *right insert* microfabricated solid support (pillars) in channel. (Losey et al. 2001, 2002). **c** Photograph of a three-layer, 20-channel/layer gas–liquid microreactor with integrated waveguides for gas–liquid flow sensing (de Mas et al. 2005)

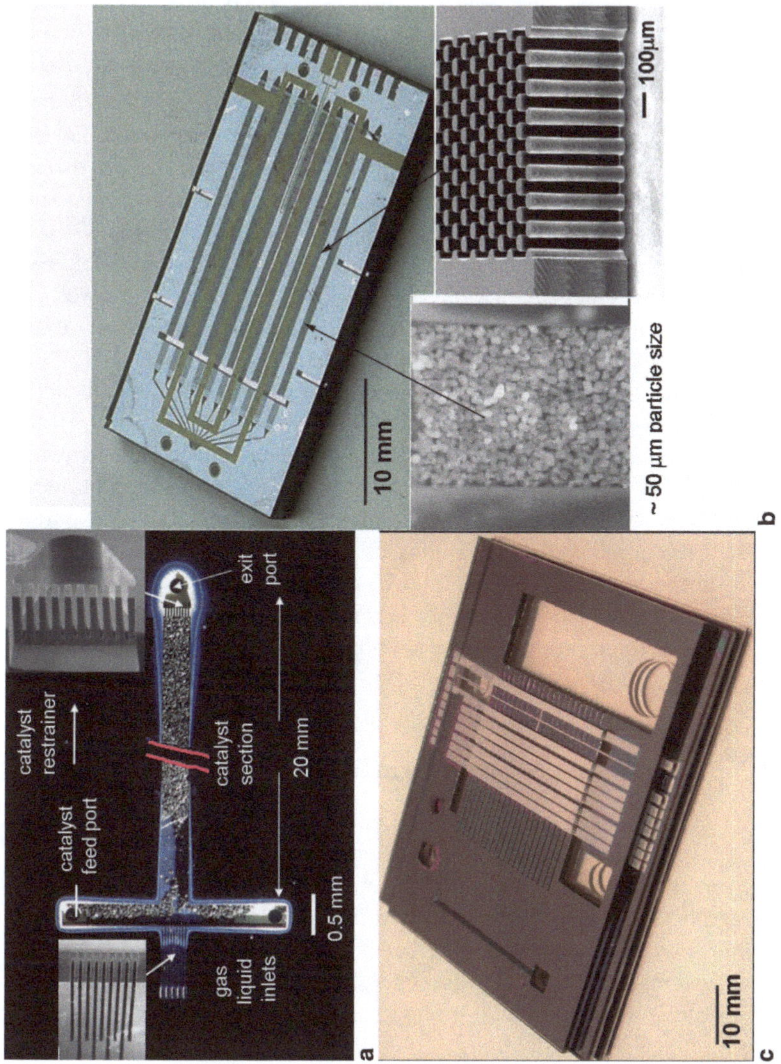
catalyst restrainer
catalyst feed port
gas
liquid
inlets
0.5 mm
catalyst section
20 mm
exit port
a
10 mm
~ 50 µm particle size
100µm
b
10 mm
c

Fig. 3. *Left* pressurized microreactor setup for aminocarbonylation of 4-bromobenzonitrile. *Right* product ratio of α-ketoamide to amide; each data point represents individual experiments (Murphy et al. 2007)

►

ble reaction conditions beyond standard bench-top equipment. For example, microreactors safely perform the direct synthesis of hydrogen peroxide from explosive mixtures of hydrogen and oxygen over a supported Pd catalyst at 20 bar (Inoue et al. 2007).

Pressurized microreactor systems greatly expand the range of reaction conditions and accelerate gas–liquid mass transfer. Moreover, applying pressure allows the use of standard solvents (e.g., toluene) above their normal boiling points (Murphy et al. 2007; Hessel et al. 2005). Reactions performed under these conditions are one of the major benefits of microwave synthesis (Kappe 2004). Microreactors offer many of the positive features of microwave reactors without requiring a microwave generator as well as providing the additional advantages of continuous flow. Heck aminocarbonylation reactions (Fig. 3) exemplify the potential for the quick and safe scanning of reagents and reaction conditions outside the scope of conventional bench-top equipment. The yield of amide increases with an increase in temperature, whereas the selectivity for α-ketoamide production is enhanced at lower temperature and higher pressure.

4 Continuous Separation in Microsystems

Continuous in-line microfluidic separators (work-up steps) are necessary for realizing continuous, multistep microchemical synthesis. Conventional techniques of extraction and distillation rely on gravity to bring about the separation; gravity drives the denser phase against the lighter phase and settles it as the bottom layer. However, in microfluidic systems surface tension forces exceed gravitational forces by several orders of magnitude, implying that droplets in one phase will remain pinned by surface tension (Günther and Jensen 2006). As a consequence, most microfluidic separation techniques cannot use gravity but must exploit surface tension effects to drive the separation. The

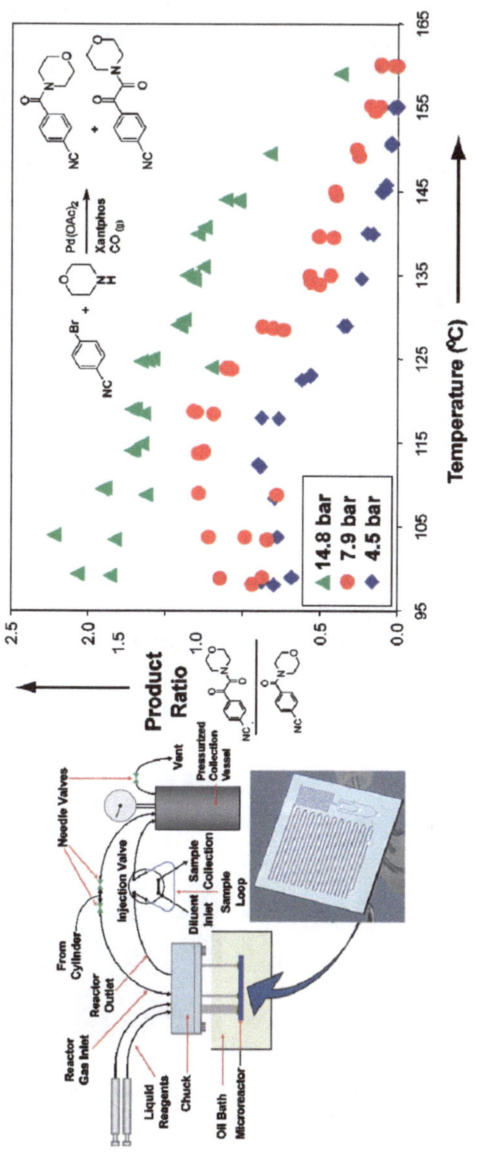
Needle Valves
Vent
Pressurized Collection Vessel
Injection Valve
Sample Collection
Diluent Inlet
Sample Loop
From Cylinder
Reactor Outlet
Reactor Gas Inlet
Liquid Reagents
Chuck
Oil Bath
Microreactor
Product Ratio
Temperature (°C)
14.8 bar
7.9 bar
4.5 bar
Pd(OAc)2
Xantphos
CO (g)

laminar flow characteristics of microfluidics are often used to implement side-by-side contacting of immiscible fluids (e.g., aqueous and organic streams) in a co-current arrangement (Robins et al. 1997; Tokeshi et al. 2000, 2002). Preserving phase separation is usually achieved by using small interfacial areas to maintain sufficient capillary pressure to counterbalance the imposed driving pressure, or by modifying channel surface wetting characteristics to stabilize interfaces. Consequently, the resulting microextraction system often has relatively low interfacial surface area to microchannel volume ratios, with corresponding modest separation throughput.

Higher separation capacity can be achieved with porous membrane separation systems in which selective wetting and capillary pressures are used to separate two fluid phases with different wetting characteristics (Fig. 4) (Günther et al. 2005; Kralj et al. 2007). In these systems, one fluid selectively wets the porous membrane, for example the organic phase in the case of a fluoropolymer, preventing the aqueous phase from penetrating. An imposed pressure, not exceeding the capillary pressure, drives the organic phase through the membrane, resulting in a quantitative separation of the two streams (Kralj et al. 2007). A similar principle applies to gas–liquid separation where the wetting liquid blocks the holes preventing the gas from crossing the membrane and effecting a quantitative separation of the gas–liquid stream. As in the case of

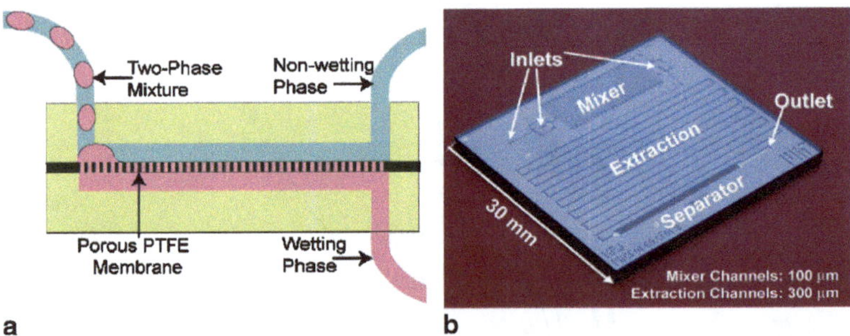

Fig. 4. **a** Schematic of porous membrane-based separation of immiscible liquids with different wetting characteristics. **b** Liquid–liquid extraction device (Kralj et al. 2007)

liquid–liquid separation, the effectiveness of the separation depends on appropriate balancing of imposed pressure and surface tension forces (Günther et al. 2005).

The separation methods allow the continuous removal of unreacted reagents and byproducts, making it possible to realize a series of transformations without leaving the microreactor environment, as exemplified by the synthesis of carbamates using the Curtius rearrangement of isocyanates (Govindan 2002; Sahoo et al. 2006). The multistep reaction sequence (Fig. 5) involved three transformations and two separations: (i) a phase transfer reaction between a aqueous azide solution and benzoyl chloride in toluene, followed by a liquid–liquid separation step to remove the aqueous stream; (ii) the rearrangement of benzoyl azide to the isocyanate by heating, followed by separation of the evolved gas (N_2); and (iii) the final liquid–liquid reaction of benzoyl isocyanate with ethanol. Each step was run to completion, and the increased contacting area per unit volume in the microreactor eliminated the need for any

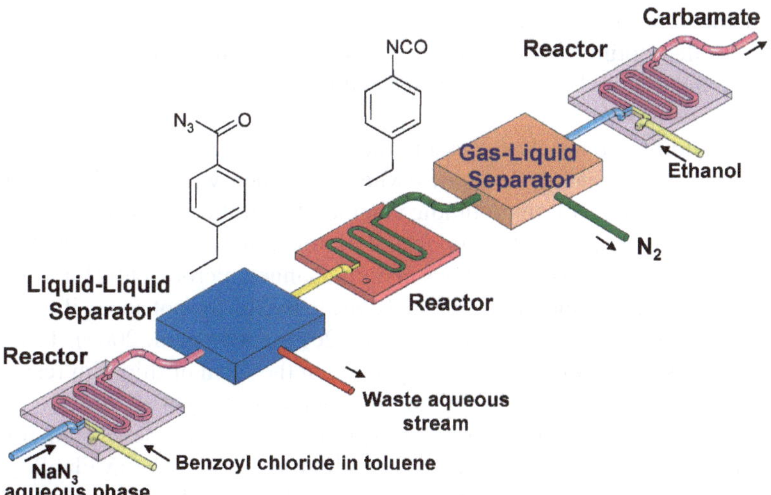

Fig. 5. System of three microreactors (Fig. 1) and two separators (Fig. 4) for continuous synthesis of carbamates from NaN_3 and organic chlorides (Sahoo et al. 2007)

phase transfer catalyst in the first step. The conversion of benzoyl azide to phenyl isocyanate in the second reactor was rate limiting and a function of the heating temperature with 99% yield at 105°C and 1 h residence time. This residence time could be further reduced by loading the second reactor with a solid acid catalyzing the rearrangement. The model chemistry also demonstrates the in situ generation and the use of organic azides and isocyanates with associated safety advantages. The multistep scheme also opens the opportunity for synthesis of chemical libraries by splitting the formed azide and isocyante streams into additional streams and reacting them with other reagents.

5 Integration of Physical and Chemical Sensors

Integrating chemical analysis methods and physical sensors with microreactors enables monitoring of reaction conditions and composition. This ability renders instrumented microreactors powerful tools for determining chemical kinetics and identifying optimal conditions for chemical reactions. The latter can be achieved by automated feedback-controlled optimization of reaction conditions, which greatly reduces time and materials costs associated with the development of chemical synthesis procedures.

In addition to absolute pressure measurements, pressure sensors can be used to determine flow rates when combined with a well-defined pressure drop over a microfluidic channel. Integration of optical wave-guide structures provides opportunities for monitoring of segmented gas–liquid or liquid–liquid flows in multichannel microreactors for multiphase reactions, including channels inside the device not accessible by conventional microscopy imaging (Fig. 2c) (de Mas et al. 2005). Temperature sensors are readily incorporated in the form of thin film resistors or simply by attaching thin thermocouples (Losey et al. 2001).

Miniaturized chemical analysis systems have been developed for most macroscopic counterparts (Dittrich et al. 2006). The availability of optical fibers, light sources, and detectors in the visible UV and near-infrared (NIR) wavelengths makes it possible to integrate spectroscopic measurements in microreactors (Löbbecke et al. 2005). Fourier transform infrared spectroscopy (FTIR) is an efficient, broadly applicable spectroscopy technique, but FTIR window materials are not compatible with microfabrication techniques and fibers are generally not available. Fortunately, silicon is transparent to IR radiation in most of the wavelength region of interest (4,000–1,000 cm^{-1}), allowing the use of silicon microreactors in a FTIR sample compartment (Floyd et al. 2005). Alternatively, the silicon can be made into a crystal for attenuated total reflection (ATR) (multiple internal reflection) FTIR and integrated into microreactors for in situ reaction monitoring (Herzig-Marx et al. 2004). Raman spectroscopy has also been incorporated as an effective monitoring tool through the use of sampling systems (Löbbecke et al. 2005) and in-line microscopes (Leung et al. 2005)

6 Handling Solid Particles in Microreactors

The generation of solid by-products can lead to plugging microreactors if the formed solid readily aggregates. However, in many cases solid by-products, such as palladium black in Pd-catalyzed Heck aminocarbonylation (Murphy et al. 2007), can be minimized by an appropriate choice of solvent—sometimes different from the commonly used solvent in batch synthesis. Additionally, it is feasible to clean microreactors, in particular glass-coated systems, by techniques similar to those used to clean macroscopic glassware. Static microfluidic systems have been developed for protein crystallization via controlled evaporation of water through PDMS (Hansen et al. 2002; Zheng et al. 2004). By controlling sheath flows, it is possible to operate continuous crystallization in microreactors as long as the interactions among crystals are sufficiently weak to avoid agglomeration.

In cases where the solid particles form stable colloidal solutions, such as semiconductor and metal oxide nanoparticles, microreactors become powerful synthesis tools producing precisely controlled particle size distributions and surface chemistry (Khan et al. 2004; Yen et al. 2005). In a single-phase reactor, as a result of the parabolic fluid-velocity profile, fluid near the wall spends more time in the reactor than fluid in the center, resulting in different reaction times. Introducing inert gas can create a segmented flow eliminating this dispersion problem by making each fluid segment a nearly independent "well-mixed beaker" owing

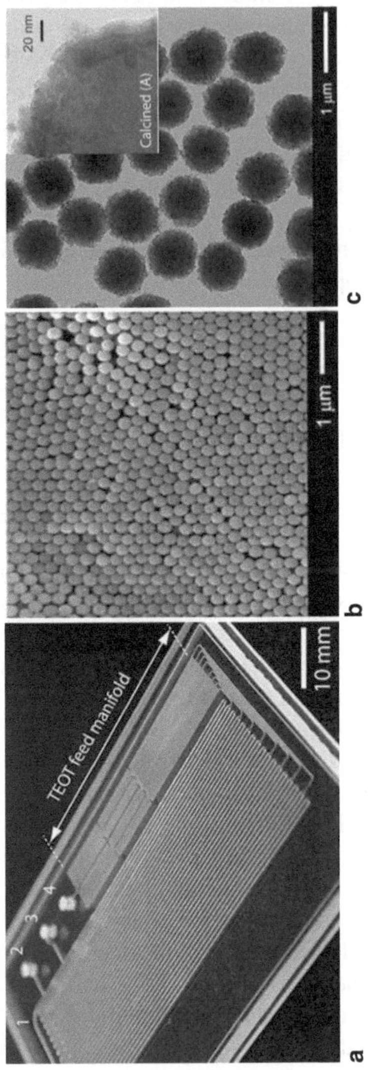
a
b
c
TEOT feed manifold
10 mm
1 μm
1 μm
Calcined (A)
20 nm
1
2
3
4

to fluid recirculations within the segment (Günther and Jensen 2006). Hence, particles with narrow distributions are obtained. Moreover, the microreactor enables the synthesis of particles with predefined sizes, and the subsequent modification of the particle surfaces in situ with organic or inorganic materials, as required by the final functionality. For example, Fig. 6 illustrates a PDMS microfluidic device used to synthesize silica particles as well as titania-silica core shell structures (Khan 2006). Gas–liquid segmented flows provided a narrow size distribution. Side inlets enabled precise control of the thickness of the titania shell without secondary nucleation of titania particles (Fig. 6b,c), a complication otherwise often experienced in batch synthesis.

When successive steps of core and shell growth or surface modifications involve similar chemistry, the microfluidic systems can be linked, but multistep processing with incompatible chemistry requires intermediate particle separation and solvent change between the different process steps. Microfluidic systems allow the integration of electrical field-driven separation processes providing rapid switching of colloidal particles from a reaction stream (containing unreacted precursors and/or other undesired species) to another liquid stream co-flowing parallel to the first (Khan 2006). Such systems enable particle washing, switching of reactants for multiple shells, and surface functionalization without the laborious ultra-centrifugation and decanting associated with typical batch processes.

7 Summary and Outlook

The microreactors described above are primarily laboratory tools intended to accelerate discovery and development. They will augment existing tools and be particularly useful in enabling highly reactive systems and challenging operating conditions (high pressures and temperatures). The continuous flow inherent in microreactor technology pro-

◄

Fig. 6. **a** PDMS device for sol-gel synthesis of core shell SiO_2–TiO_2 colloids. **b** SEM image of synthesized colloids. **c** TEM image of colloids; TiO_2 is shell-visible (Khan 2006)

vides opportunities for faster scaling of chemistry from laboratory to production as well as the on-demand, on-site production of chemicals with storage or shipping limitations. Scaling to moderate levels of production (a few metric tons per year) can be accomplished, in principle, by simply numbering up the microreactors. However, in many cases it will be more advantageous to use instrumented microreactors to obtain the information on chemical kinetics and optimal operating conditions from microreactors to design continuous reactor systems capable of producing hundreds of tons of product per year. Advances in computational fluid dynamics and reactor design imply that the rate-limiting step in scale-up is often inadequate chemical information. Furthermore, experienced gained with analytical methods on the microreactor scale facilitates the implementation of process analytical techniques for product quality control.

The examples given above as well as those in other chapters of this volume demonstrate the utility of microreaction technology in chemical research and development. However, the development of cost-effective, integrated, easy to operate microchemical systems remains a major challenge to realizing the speed and cost advantages promised by chemical microsystems.

Acknowledgements. The author thanks members of the microreactor research group and collaborators for the work forming the basis for this summary, J. Mc-Mullen and N. Zaborenko for critical reading of the manuscript, and DARPA, ARO-MURI and the MicroChemical Systems Technology Center for funding.

References

Baxendale IR, Deeley J, Griffiths-Jones CM, Ley SV, Saaby S, Tranmer GK (2006) A flow process for the multi-step synthesis of the alkaloid natural product oxomaritidine: a new paradigm for molecular assembly. Chem Commun, p 2566–2568

Chambers RD, Spink RCH (1999) Microreactors for elemental fluorine. Chem Commun, p 883–884

De Mas N, Günther A, Schmidt MA, Jensen KF (2003) Microfabricated multiphase reactors for the selective direct fluorination of aromatics. Ind Eng Chem Res 42:698–710

De Mas N, Günther A, Kraus T, Schmidt MA, Jensen KF (2005) Scaled-out multilayer gas–liquid microreactor with integrated velocimetry sensors. Ind Eng Chem Res 44:8997–9013

Dittrich PS, Tachikawa K, Manz A (2006) Micro total analysis systems. Latest advancements and trends. Anal Chem 78:3887–3907

Ehrfeld W (1997) (ed) Microreaction technology. Springer, Berlin New York Heidelberg

Ehrfeld W, Golbig K, Hessel V, Lowe H, Richter T (1999) Characterization of mixing in micromixers by a test reaction: single mixing units and mixer arrays. Ind Eng Chem Res 38:1075–1082

Ehrfeld W, Hessel V, Lowe H (2000) Microreactors: new technology for modern chemistry. Wiley-VCH, Weinheim, Germany

Fletcher PDI, Haswell SJ, Pombo-Villar E, Warrington BH, Watts P, Wong SYF, Zhang XL (2002) Microreactors: principles and applications in organic synthesis. Tetrahedron 58:4735–4757

Floyd TM, Schmidt MA, Jensen KF (2005) Silicon micromixers with infrared detection for studies of liquid-phase reactions. Ind Eng Chem Res 44:2351–2358

Fredrickson CK, Fan ZH (2004) Macro-to-micro interfaces for microfluidic devices. Lab Chip 4:526–533

Govindan CK (2002) An improved process for the preparation of benzyl-n-vinyl carbamate. Org Process Res Dev 6:74–77

Günther A, Jensen KF (2006) Multiphase microfluidics: from flow characteristics to chemical and materials synthesis. Lab Chip 6:1487–1503

Günther A, Jhunjhunwala M, Thalmann M, Schmidt MA, Jensen KF (2005) Micromixing of miscible liquids in segmented gas liquid flow. Langmuir 21:1547–1555

Hansen CL, Skordalakes E, Berger JM, Quake SR (2002) A robust and scalable microfluidic metering method that allows protein crystal growth by free interface diffusion. Proc Natl Acad Sci USA 99:16531–16536

Herzig-Marx R, Queeney KT, Rebecca JJ, Schmidt MA, Jensen KF (2004) Infrared spectroscopy for chemically specific sensing in silicon-based microreactors. Anal Chem 76:6476–6483

Hessel V, Hofmann C, Löb P, Löhndorf J, Löwe H, Ziogas A (2005) Aqueous Kolbe-Schmitt synthesis using resorcinol in a microreactor laboratory rig under high-P,T conditions. Org Process Res Dev 9:479–489

Inoue T, Schmidt MA, Jensen KF (2007) Microfabricated multiphase reactors for the direct synthesis of hydrogen peroxide from hydrogen and oxygen. Ind Eng Chem Res 46:1153–1160

Jackman RJ, Floyd TM, Ghodssi R, Schmidt MA, Jensen KF (2001) Microfluidic systems with on-line UV detection fabricated in photodefinable epoxy. J Micromech Microeng 11:263–269

Jähnisch K, Baerns M, Hessel V, Ehrfeld W, Haverkamp V, Lowe H, Wille C, Guber A (2000) Direct fluorination of toluene using elemental fluorine in gas/liquid microreactors. J Fluorine Chem 105:117–128

Jähnisch K, Hessel V, Lowe H, Baerns M (2004) Chemistry in microstructured reactors. Angew Chem Int Ed 43:406–446

Kappe CO (2004) Controlled microwave heating in modern organic synthesis. Angew Chem Int Ed 43:6250–6284

Khan SA (2006) Microfluidic synthesis of colloidal nanomaterials. PhD dissertation, Massachusetts Institute of Technology, Cambridge

Khan SA, Gunther A, Schmidt MA, Jensen KF (2004) Microfluidic synthesis of colloidal silica. Langmuir 20:8604–8611

Kikutani Y, Hibara A, Uchiyama K, Hisamoto H, Tokeshi M, Kitamori T (2002) Pile-up glass microreactor. Lab Chip 2:193–196

Knitter R, Gohring D, Risthaus P, Hausselt J (2001) Microfabrication of ceramic microreactors Microsys Technol 7:85–90

Kobayashi J, Mori Y, Okamoto K, Akiyama R, Ueno M, Kitamori T, Kobayashi S (2004) A microfluidic device for conducting gas-liquid-solid hydrogenation reactions. Science 304:1305–1308

Kralj JG, Sahoo HR, Jensen KF (2007) Integrated continuous microfluidic liquid–liquid extraction. Lab Chip 7:256–263

Lee CC, Sui GD, Elizarov A, Shu CYJ, Shin YS, Dooley AN, Huang J, Daridon A, Wyatt P, Stout D, Kolb HC, Witte ON, Satyamurthy N, Heath JR, Phelps ME, Quake SR, Tseng HR (2005) Multistep synthesis of a radiolabeled imaging probe using integrated microfluidics. Science 310:1793–1796

Lee JN, Park C, Whitesides GM (2003) Solvent compatibility of poly(dimethylsiloxane)-based microfluidic devices. Anal Chem 75:6544–6554

Leung S-A, Winkle RF, Wootton RCR, deMello AJ (2005) A method for rapid reaction optimisation in continuous-flow microfluidic reactors using online raman spectroscopic detection. Analyst 130:46–51

Löbbecke S, Ferstl W, Pani S, Türcke T (2005) Concepts for modularization and automation of microreaction technology. Chem Eng Tech 28:484–493

London AP, Ayon AA, Epstein AH, Spearing SM, Harrison T, Peles Y, Kerrebrock JL (2001) Microfabrication of a high-pressure bipropellant microrocket engine. Sensors Actuators A 92:351–357

Losey MW, Schmidt MA, Jensen KF (2001) Microfabricated multiphase packed-bed reactors: characterization of mass transfer and reactions. Ind Eng Chem Res 40:2555–2562

Losey MW, Jackman RJ, Firebaugh SL, Schmidt MA, Jensen KF (2002) Design and fabrication of microfluidic devices for multiphase mixing and reaction. J Microelectromech Syst 11:709–717

Madou MJ (2002) Fundamentals of microfabrication: the science of miniaturization. CRC Press, Boca Raton

Murphy ER (2006) Microchemical systems for rapid optimization of organic synthesis. Ph.D. Thesis, Massachusetts Institute of Technology, Cambridge

Murphy ER, Martinelli JR, Zaborenko N, Buchwald SL, Jensen KF (2007) Accelerating reactions with microreactors at elevated temperatures and pressures: profiling aminocarbonylation reactions. Angew Chem Int Ed 46:1734–1737

Ottino JM, Wiggins S (2004) Introduction: mixing in microfluidics. Philos Trans R Soc London, A 362:923–935

Pennemann H, Watts P, Haswell SJ, Hessel V, Lowe H (2004) Benchmarking of microreactor applications. Org Process Res Dev 8:422–439

Ratner DM, Murphy ER, Jhunjhunwala M, Snyder DA, Jensen KF, Seeberger PH (2005) Microreactor-based reaction optimization in organic chemistry—glycosylation as a challenge. Chem Commun 5:578–580

Robins I, Shaw J, Miller B, Turner C, Harper M (1997) Solute transfer by liquid/liquid exchange without mixing in micro-contactor devices. In: Ehrfeld W (ed) Microreaction technology: proceedings of the first international conference on microreaction technology. Springer, Berlin New York Heidelberg, pp 35–46

Sahoo HR, Kralj JG, Jensen KF (2007) Multi-step continuous flow microchemical synthesis involving multiple reactions and separations. Angew Chem Int Ed, DOI anie.200701434

Song H, Tice JD, Ismagilov RF (2003) A microfluidic system for controlling reaction networks in time. Angew Chem Int Ed 42:768–772

Stroock AD, Dertinger SKW, Ajdari A, Mezic I, Stone HA, Whitesides GM (2002) Chaotic mixer for microchannels. Science 295:647–651

Tokeshi M, Minagawa T, Kitamori T (2000) Integration of a microextraction system on a glass chip: ion-pair solvent extraction of fe(ii) with 4,7-diphenyl-1,10-phenanthrolinedisulfonic acid and tri-n-octylmethylammonium chloride. Anal Chem 72:1711–1714

Tokeshi M, Minagawa T, Uchiyama K, Hibara A, Sato K, Hisamoto H, Kitamori T (2002) Continuous-flow chemical processing on a microchip by combining microunit operations and a multiphase flow network. Anal Chem 74:1565–1571

Wada Y, Schmidt MA, Jensen KF (2006) Flow distribution and ozonolysis in gas-liquid multichannel microreactors. Ind Eng Chem Res 45:8036–8042

Xia YN, Whitesides GM (1998) Soft lithography. Angew Chem Int Ed 37:551–575

Yen BKH, Gunther A, Schmidt MA, Jensen KF, Bawendi MG (2005) A microfabricated gas-liquid segmented flow reactor for high-temperature synthesis: the case of CdSe quantum dots. Angew Chem Int Ed 44:5447–5451

Zheng B, Tice JD, Roach LS, Ismagilov RF (2004) A droplet-based, composite pdms/glass capillary microfluidic system for evaluating protein crystallization conditions by microbatch and vapor-diffusion methods with on-chip x-ray diffraction. Angew Chem Int Ed 43:2508–2511

Ernst Schering Foundation Symposium Proceedings, Vol. 3, pp. 77–98
DOI 10.1007/2789_2007_029

Published Online: 23 May 2007

Isoindolones and Related N-Heterocycles via Palladium Nanoparticle-Catalyzed 3-Component Cascade Reactions

R. Grigg, V. Sridharan(✉)

Chemistry Department, Leeds University, LS2 9JT Leeds, UK
email: *V.Sridharan@leeds.ac.uk*

Abstract. Non-phosphine-containing cyclopalladated *N*-heterocycles possessing either sp^2 C–Pd(II) or sp^3 C–Pd(II) bonds and simple Pd(II) salts are precursors of Pd(0) nanoparticles whose initial morphology is dependent on the nature of the precursor. Addition of polyvinylpyrrolidone (pvp) dramatically increases catalyst lifetime. Nanoparticle generation can be achieved at ambient temperature in the presence of carbon monoxide by a process akin to the water–gas shift reaction. Allene also lowers the temperature required for nanoparticle generation. 3-Component catalytic cascades employing one or both of these substrates provide access to a variety of 5- and 6-membered N-heterocycles including isoindolones, *N*-aminoisoindolones, phthalazones, dihydroisoquinolines, and isoquinolones.

1 Introduction

The area of palladium catalyzed processes continues to display tremendous vitality with growth in new reactions, new ligands, and ongoing refinement of mechanistic insights (Amatore and Jutand 2000; Beletskaya and Cheprakov 2000; Evans et al. 2005). A significant factor in these advances is the key role of ligandless and ligated palladium (0) nanoparticles dating from the seminal papers of Jeffery (Jeffery 1984, 1996), Herrmann (Herrmann et al. 2005), and Milstein (Ohff et al. 1999; Weissman and Milstein 1999). The realization that cyclometallated Pd (II) complexes (Dupont et al. 2005) are, upon thermal activation, precursors of ligated Pd(0) nanoparticles has driven the development of a diverse range of palladacycles (PdCy), including cyclometallated phosphines (Herrmann et al. 1995), phosphites (Albisson et al. 1998; Bedford et al. 2002a,b), carbenes (Palencia et al. 2004), imines (Ohff et al. 1999; Weissman and Milstein 1999), heterocycles (Gai et al. 2000b), thioether (Gruber et al. 2000; Munoz et al. 2001), and oximes (Iyer et al. 2000, 2001; Alonso et al. 2002).

These precatalysts invariably contain sp^2 carbon–palladium bonds and usually require temperatures in excess of 120°C for their transformation into the active Pd(0) nanoparticle catalysts.

Our interest in this general area focused on four aspects of nonphosphine PdCys, namely:

1. A comparison of sp^2 and sp^3 carbon–palladium-bonded PdCys incorporating heterocycles
2. The effect of substituents in the heterocycle in controlling release/reactivity of the Pd(0) nanoparticles
3. Other nonthermal methods of activating the release of the Pd(0) nanoparticles
4. Application of the PdCy precatalysts to multicomponent reactions (MCRs)

2 Synthesis of Palladacycles

The PdCys were synthesized by heating the appropriate ligand with palladium acetate in acetic acid (Scheme 1). In each case, they comprised a mixture of *cis* and *trans* isomers whose ratio varied with the ligand.

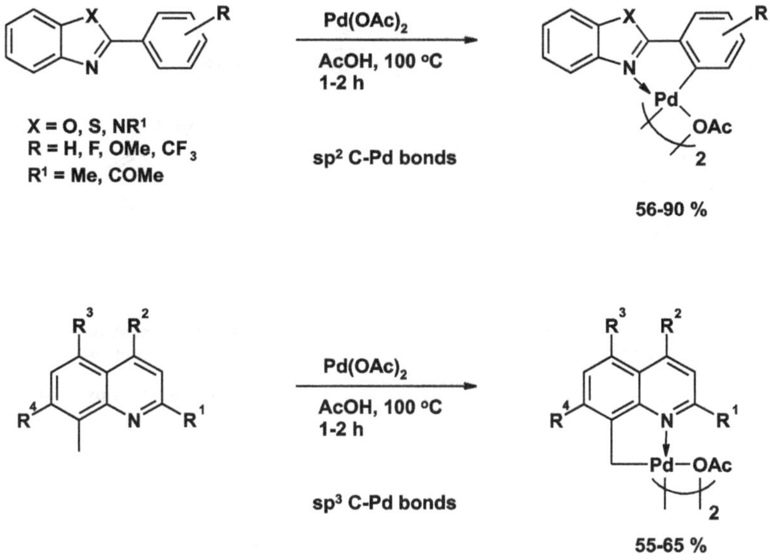

X = O, S, NR^1
R = H, F, OMe, CF_3
R^1 = Me, COMe

$Pd(OAc)_2$
AcOH, 100 °C
1-2 h

sp^2 C-Pd bonds

56-90 %

$Pd(OAc)_2$
AcOH, 100 °C
1-2 h

sp^3 C-Pd bonds

55-65 %

Scheme 1.

The isomers were separable by chromatography but were generally used as mixtures.

3 Formation of Palladium Nanoparticles: Heck Reaction

A standard Heck reaction was used to identify the most promising PdCys in each series (Scheme 2) (Gai et al. 2000b); Evans et al. 2005).

Several factors were apparent from our initial screens:

1. Addition of pvp significantly extends catalyst lifetime (Scheme 2). This is a well-known protocol whereby the Pd nanoparticles are wrapped up in the polymer chains, which prevents their aggregation into catalytically inactive palladium black.
2. The PdCy precatalysts showed zero or very low activity with ArCl substrates.

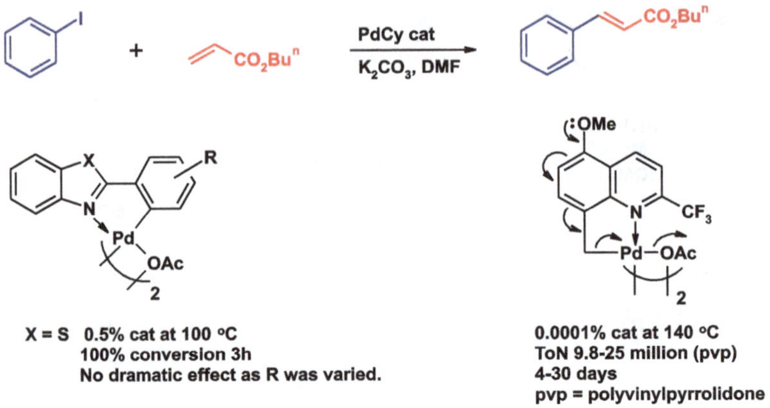
PdCy cat
K_2CO_3, DMF
CO_2Bu^n
CO_2Bu^n
X
R
N
Pd
OAc
2
:OMe
N
CF_3
Pd—OAc
2
X = S 0.5% cat at 100 °C
100% conversion 3h
No dramatic effect as R was varied.
0.0001% cat at 140 °C
ToN 9.8-25 million (pvp)
4-30 days
pvp = polyvinylpyrrolidone

Scheme 2.

3. There was an induction period whose length depended on reaction temperature. Further evaluation of the PdCy precatalysts showed that the reaction temperatures required were significantly lower for processes involving reactions of ArPdX with allenes (Schemes 3 and 4).

In Scheme 3 we see an interesting reversal or umpolung of the normal π-allylpalladium(II) chemistry. π-Allyl palladium(II) complexes are electrophilic reacting, for example, with amines, alcohols, carbanions, etc. In Scheme 3, the π-allyl intermediate displays a nucleophilic character. This effect is very sensitive to the size of the ring being formed and the metal carbonate cation and it is not observed in intermolecular reactions (Gai et al. 2000a). Moreover, the analogous six-membered ring cyclization afforded only 34% of product. We have developed a much more synthetically flexible Pd(0)/In(0) bimetallic cascade for achieving both intra- and intermolecular umpolung processes of π-allyl species (Anwar et al. 2000; Cooper et al. 2002a,b; Cleghorn et al. 2003a,b, 2005).

A related process (Scheme 4) (Gai et al. 2000b; Evans et al. 2005) also occurs at moderate temperatures.

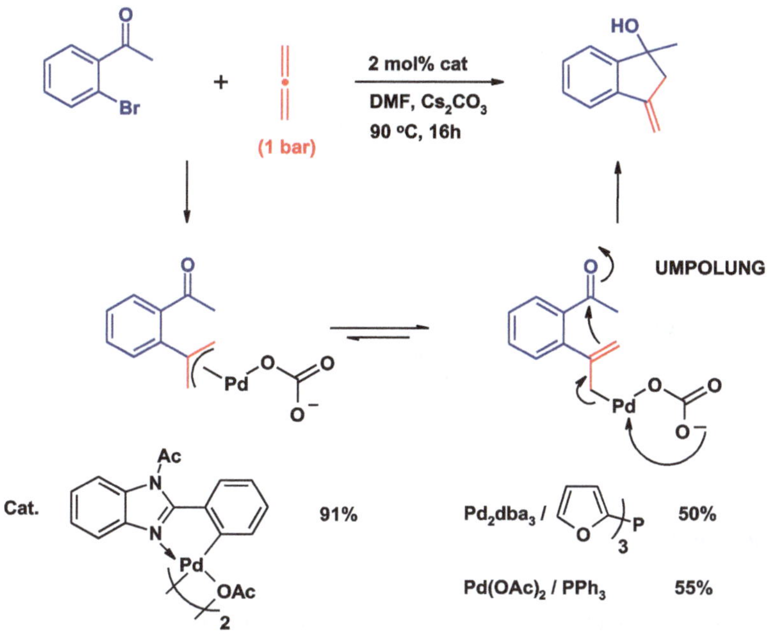

Scheme 3.

4 Formation of Palladium Nanoparticles in the Presence of CO

It must be noted that the reactions shown in Schemes 3 and 4 have not been optimized. An insightful discussion of the effects of solvent and phosphine on the formation of coordinatively unsaturated species from Pd_2dba_3/R_3P has been given by Amatore and Jutand (1998).

A key feature of PdCys as precursors of Pd(0) nanoparticles is that reduction of Pd(II) → Pd(0) involving C–Pd bond cleavage is required. This accounts for both the high temperatures invariably required and the induction period in the absence of reductants. Rosner et al. have developed a detailed kinetic model of a Heck reaction catalyzed by dimeric palladacycles (Rosner et al. 2001a,b). This model explains the experimental observations and is consistent with an active species

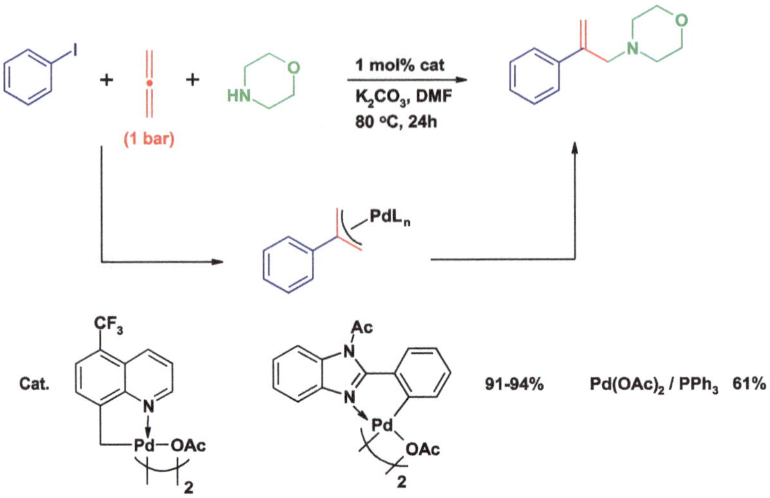

Scheme 4.

being slowly metered into the reaction. Comparison between phosphine and non-phosphine-based palladacycles suggests that they follow the same reaction mechanism. They have also highlighted the role of water in accelerating the formation of the active catalyst species. Thus, the rate-determining metering step is outside the true catalytic cycle, in the case of aryl iodides and activated aryl bromides, and this has important consequences for the use of these catalysts. This, of course, does not apply to the case of unreactive aryl chlorides and bromides where oxidative addition is rate-determining. Seminal contributions to these multifactorial processes have also been made by van Srijdonck and Hartwig's groups (Louie and Hartwig 1996; van Srijdonck et al. 1999). TONs with these palladacycles would appear even more impressive if based on the actual amount of Pd present in the catalytic cycle at any one time (Amatore and Jutand 2000). In our PdCy catalysts, the function of the substituents in the PdCys can be interpreted as perturbing the sp^2 or sp^3 C–Pd covalent bond and the N–Pd dative bond and in so doing controlling the rate of release of Pd nanoparticles into solution. Reductive methods for generating a range of metal nanoparticles have been reported (Roucoux et al. 2002), some of which employ carbon monoxide (Kopple et al. 1980; Mucalo and Coouey 1989; Powell and Dahl 2000). This encouraged exploration of carbon monoxide as both reductant and reagent for PdCys (Table 1) (Grigg et al. 2004). There is a noticeable solvent effect (Table 1) on both the rate of nanoparticle formation and their stability. In toluene aggregation to Pd black and sedimentation occurs fairly rapidly. It seems likely that a major pathway for nanoparticle formation in these cases is a process akin to the water–gas shift reaction (Scheme 5) (Hermann and Muchlhofer 2002).

We investigated the fate of the PdCy ligand in several cases. Thus **1** afforded an approximately 3:1 mixture of **2** and **3** (Scheme 6) (Grigg et al. 2004).

The structure of **3** was confirmed by X-ray crystallography (Fig. 1). The morphology of the nanoparticles was examined by transmission electron microscopy (TEM). The two sp^2-C palladacycles **1** and **4** gave what appeared to be triangular nanoparticles, in 2D, from 2–12 nm in size while the sp^3-C PdCys **5** and **6** and $Pd(OAc)_2$ exhibited more conventional morphology and were faceted palladium particles from 3–10 nm (Fig. 2).

Table 1 Formation of Pd(0) nanoparticles by treatment of PdCy's with CO (1 atm at 25°C)

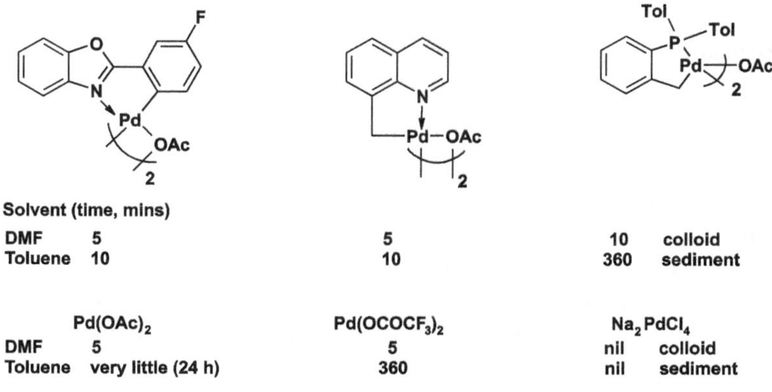

Solvent (time, mins)			
DMF	5	5	10 colloid
Toluene	10	10	360 sediment
	$Pd(OAc)_2$	$Pd(OCOCF_3)_2$	Na_2PdCl_4
DMF	5	5	nil colloid
Toluene	very little (24 h)	360	nil sediment

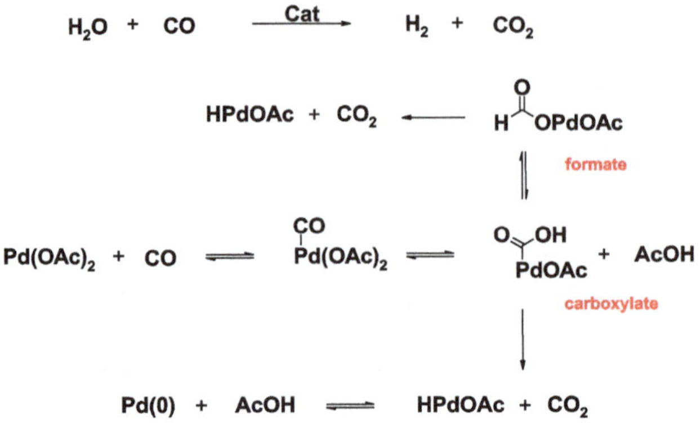

Scheme 5.

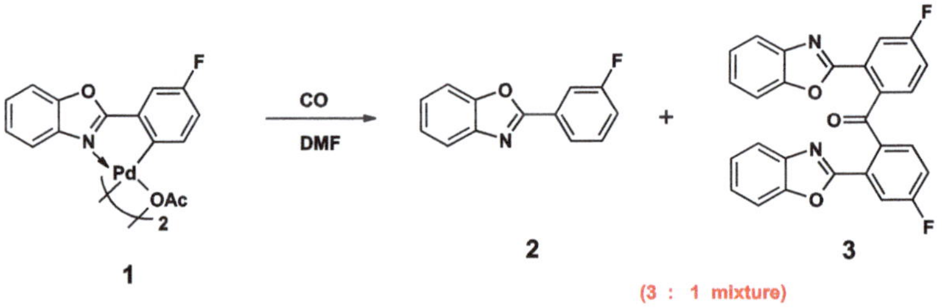

Scheme 6.

Although the nature of the PdCy influences the structure of the initially formed nanoparticles, it must be noted that once they enter the catalytic cycle this structural biasing effect will be lost.

With this rapid room temperature protocol for generating Pd(0) nanoparticles, we were encouraged to explore its application to the synthesis of isoindolinones.

The isoindolinone motif occurs a range of drugs, for example Fig. 3.

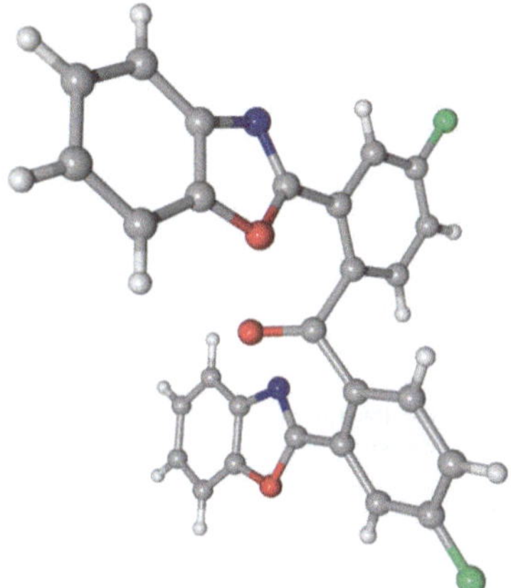

Fig. 1.

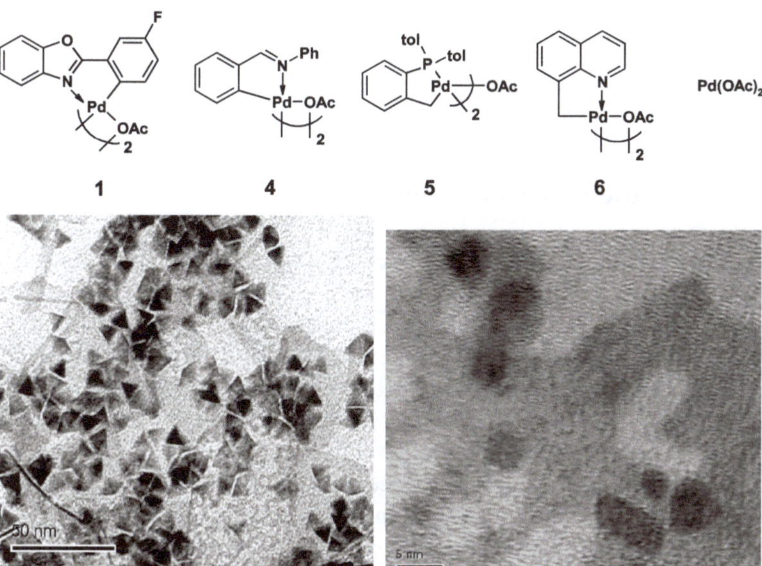

Fig. 2.

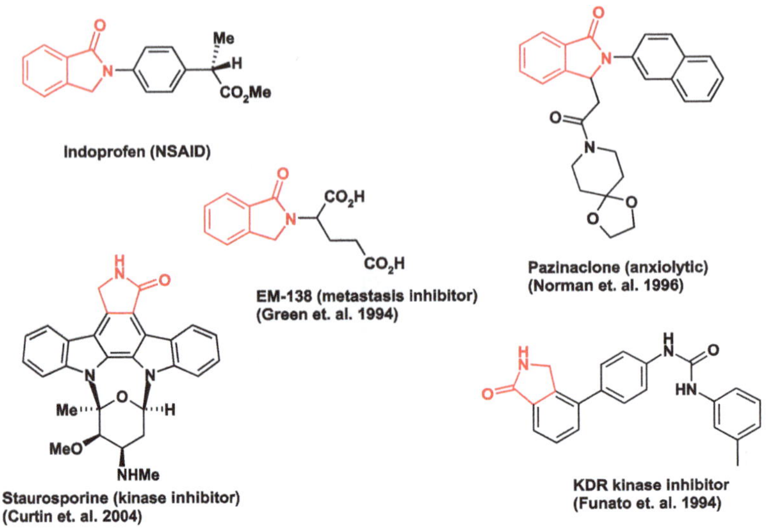

Indoprofen (NSAID)

EM-138 (metastasis inhibitor)
(Green et. al. 1994)

Pazinaclone (anxiolytic)
(Norman et. al. 1996)

Staurosporine (kinase inhibitor)
(Curtin et. al. 2004)

KDR kinase inhibitor
(Funato et. al. 1994)

Fig. 3.

5 Synthesis of Isoindolones

Initially we studied a three-component cascade (Scheme 7) at room temperature (Grigg et al. 2003).

Variation of the pK_a of the amine enabled us to show that the process occurred via path A since intermediate **8** was clearly visible in all reactions and conversion to product **10** correlated with the pK_as of the primary amines **11** and the secondary amines **8** (Table 2, Figs. 4a and b).

Table 2 shows a strong correlation between the pK_a's of amines **11** and **8** with the most basic **11a** proceeding to 98% conversion to product over 48 h at rt (Table 2, entry 1) and the least basic Table 2 (entries 5 and 6) exhibiting very low conversion to product.

When the reaction temperature is increased to 80°C, all amines convert to product in less than 10 h. Typical examples and yields (unoptimized) are shown in Fig. 5. It is interesting that the allyl amine double bond does not isomerize in the first example.

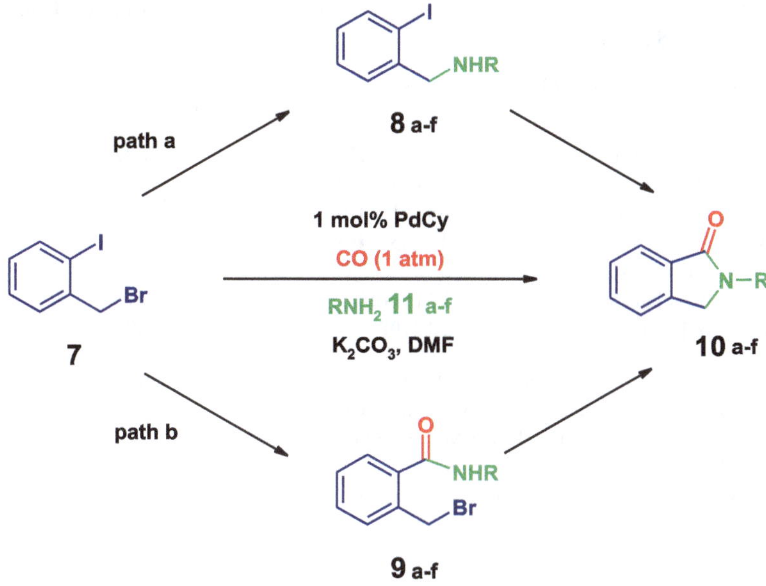

Scheme 7.

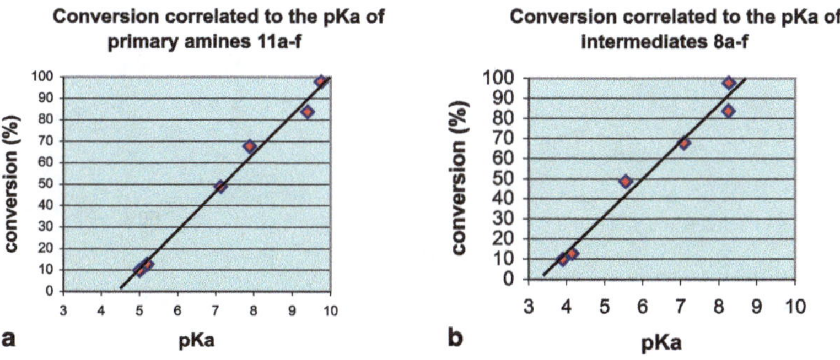

Fig. 4.

Table 2 Correlation of conversion to isoindolone with pK_a primary **11** and secondary **8** amines

Entry	Primary amine **11**	Conv.[b] (%)	pK_a^d of amines **11**	pK_a^d of amines **8**
1	R-α-Methyl benzylamine	98 (81[c], **10a**)	9.75	8.27
2	Benzylamine	84 (72[c], **10b**)	9.40	8.25
3	Propargyl amine	68 (51[c], **10c**)	7.89	7.09
4	L-Phenylalanine methyl ester	49 (33[c], **10d**)	7.13	5.56
5	p-Anisidine	13 (**10e**)	5.2	4.15
6	p-Toluidine	10 (**10f**)	5.0	3.9

[a] All reactions carried out in DMF at rt for 48 h in the presence of 1 mol% PdCy **1**, carbon monoxide (1 atm) and K_2CO_3;
[b] Conversion ddetermined by NMR based on intermediated **8**;
[c] Isolated Yield;
[d] pK_a calculated using ACD/pK_a software

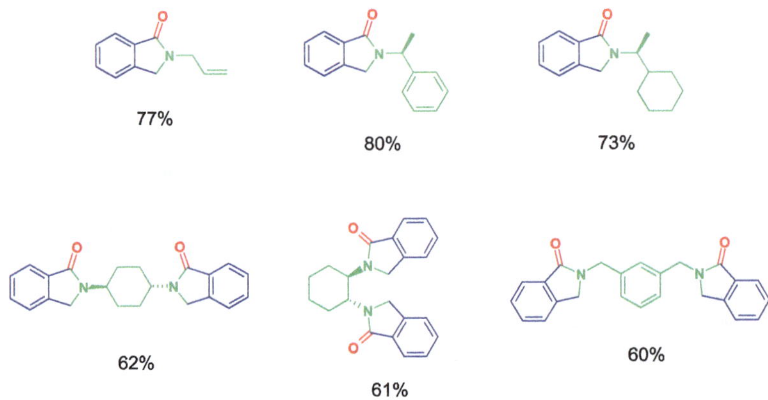

Fig. 5.

3-Substituted isoindolinones feature in a number of anxiolytic drugs (e.g., Fig. 3) and this led us to devise a general three-component cascade for accessing this pharmacophore (Scheme 8) (Gai et al. 2003b, 2005).

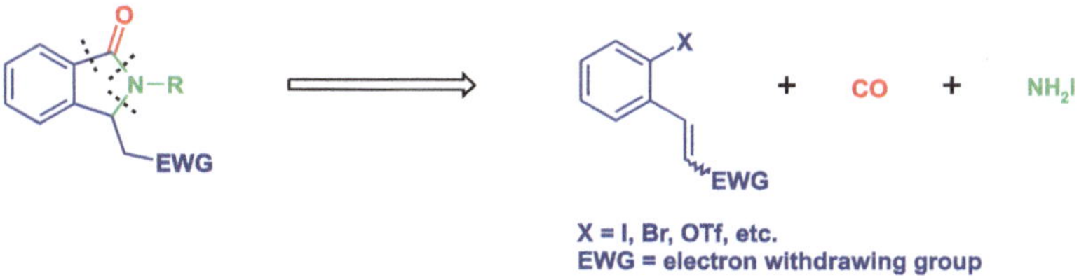

X = I, Br, OTf, etc.
EWG = electron withdrawing group

Scheme 8.

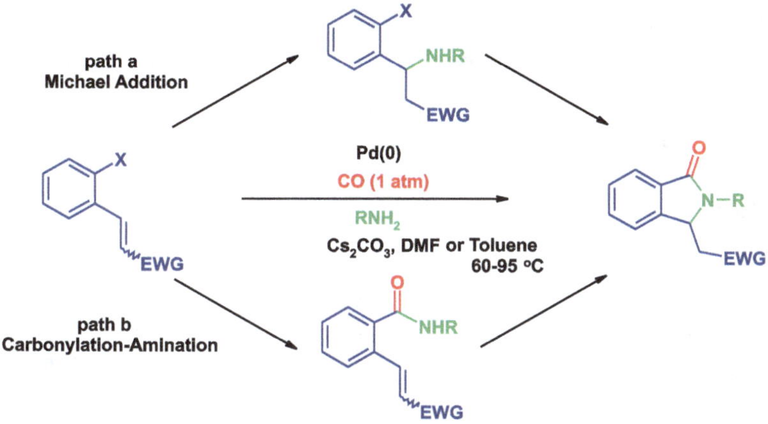

Scheme 9.

The objective was to engineer a catalytic carbonylation-amination-Michael addition cascade employing carbon monoxide (1 atm). This cascade, if viable, could, conceptually, proceed via either of two pathways (Scheme 9).

Control experiments in which carbon monoxide was omitted employing benzylamine as the nucleophile and CO_2Me as the EWG group failed to give any Michael adduct under conditions where the cascade proceeded normally in the presence of CO. Hence, in cases so far studied, we believe path A is operational and that the amide is deprotonated to provide an anionic nucleophile in the final step.

The reaction tolerates a wide variety of EWG groups (ester, amide, nitrile, ketone, sulphonyl, 2-pyridyl etc), and amines (aliphatic, aryl,

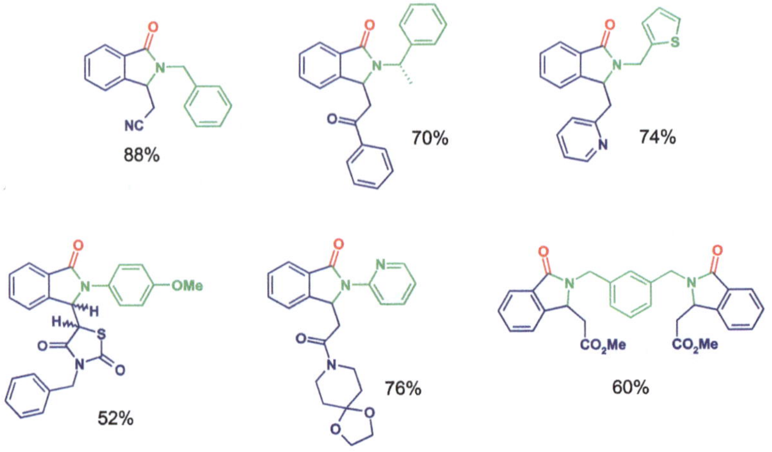

Fig. 6.

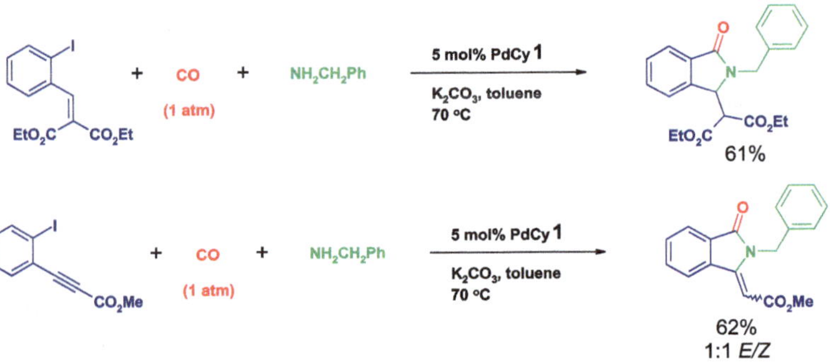

Scheme 10.

2-aminopyrimidine, 2-aminopyridine etc.). Some typical examples are shown in Fig. 6.

Significant further variation is possible including the employment of doubly activated Michael acceptors and alkynyl Michael acceptors (Scheme 10).

In the latter example in Scheme 10, the enamide moiety provides an accessible E/Z-equilibration pathway (Scheme 11).

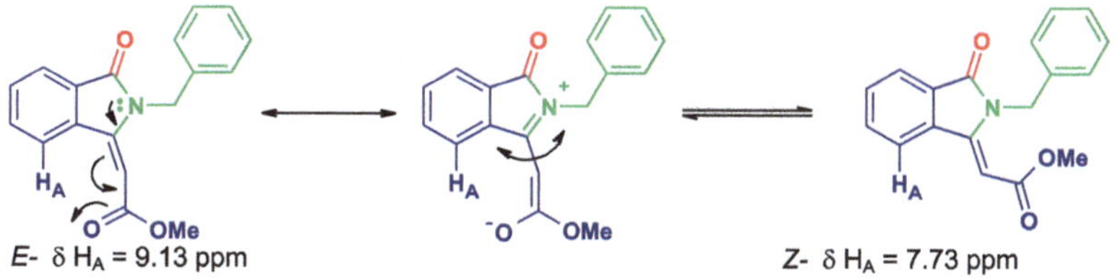

Scheme 11.

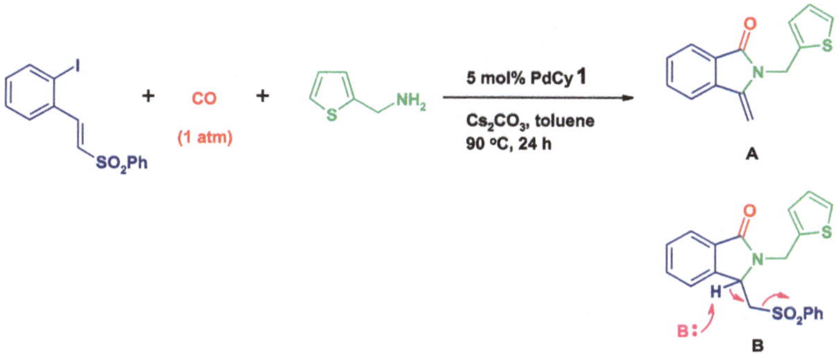

Scheme 12.

When a phenylsulphonyl EWG is used, the product depends on the reaction time (Scheme 12). Shorter times lead to mixtures of A and B.

This cascade has also been adapted to the synthesis of six-membered *N*-heterocycles by employing 1,2-disubstituted hydrazines as the nucleophile (Scheme 13) (unpublished data).

Monosubstituted hydrazines give *N*-amino isoindolones (Scheme 14) (unpublished data) with potentially useful pharmacological properties.

6 Synthesis of Tetrahydroisoquinolines

A second approach to six-membered heterocycles can be achieved by replacing carbon monoxide by allene (Scheme 15) (Gai et al. 2003a).

This cascade has been exemplified with both carbon nucleophiles and primary amine nucleophiles. We have also carried out these processes

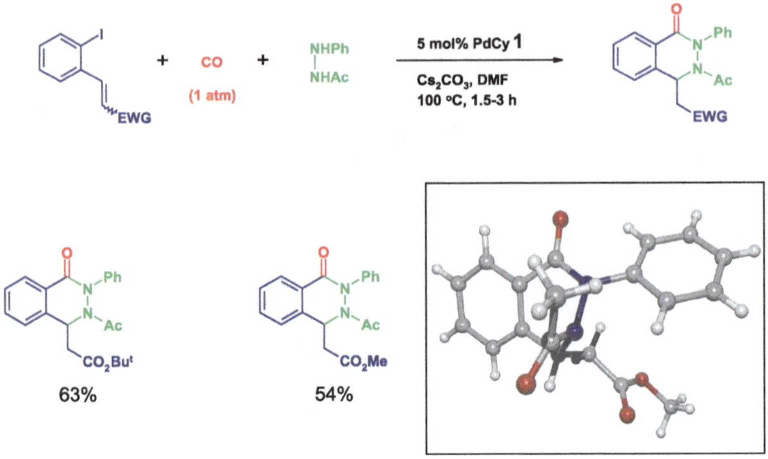

Scheme 13.

in the microwave (unpublished data). Typical examples are shown in Fig. 7.

7 Synthesis of Isoquinolinones

A complementary approach to six-membered rings can be achieved with methyl 2-iodobenzoate or 2-iodobenzoyl chloride, as shown in Scheme 16 (Grigg et al. 2002).

Less nucleophilic amines are readily incorporated into these cascades if PdCy precatalysts and acyl chlorides are employed. Typical examples are shown in Fig. 8. In these cases, the intermediate amide can be isolated.

8 Summary

PdCys are a valuable addition to the range of Pd(0) precatalysts. The thermal reductive degradation of PdCys to Pd(0) nanoparticles generally requires temperatures in excess of 110°C but is sensitive to the substrates involved in the reaction. In appropriate cases, it can be achieved

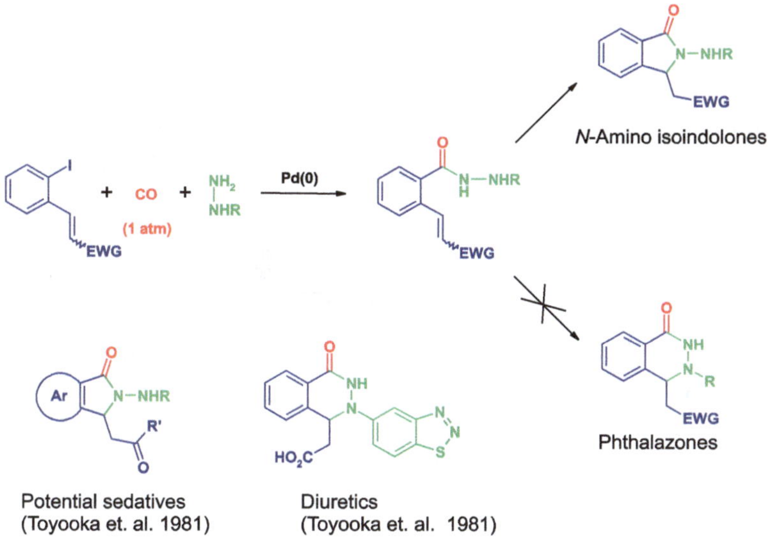

N-Amino isoindolones

Pd(0)

CO (1 atm)

Phthalazones

Potential sedatives
(Toyooka et. al. 1981)

Diuretics
(Toyooka et. al. 1981)

Scheme 14.

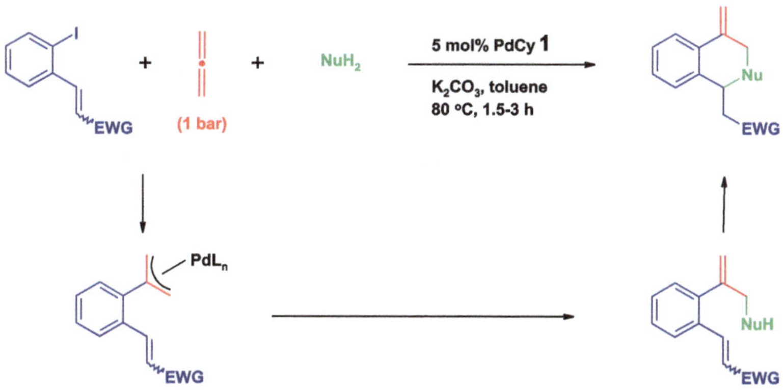

5 mol% PdCy **1**

K_2CO_3, toluene
80 °C, 1.5-3 h

(1 bar)

NuH_2

PdL_n

Scheme 15.

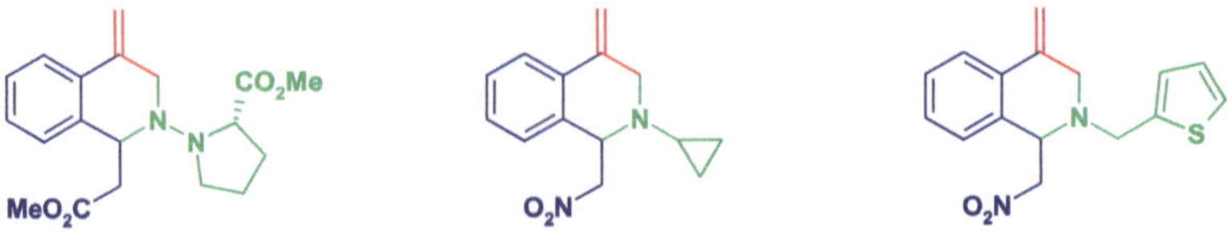

Fig. 7.

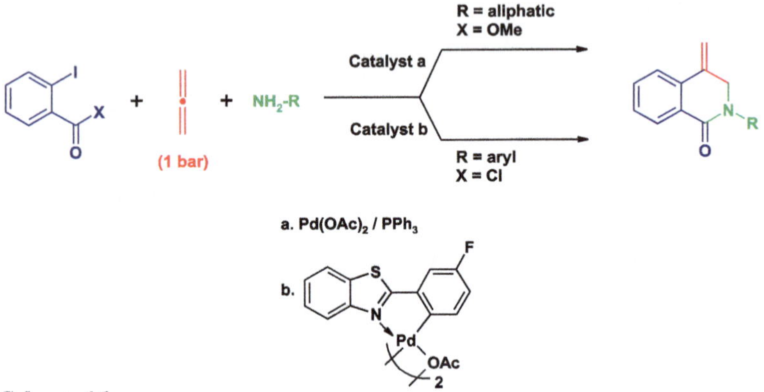

Scheme 16.

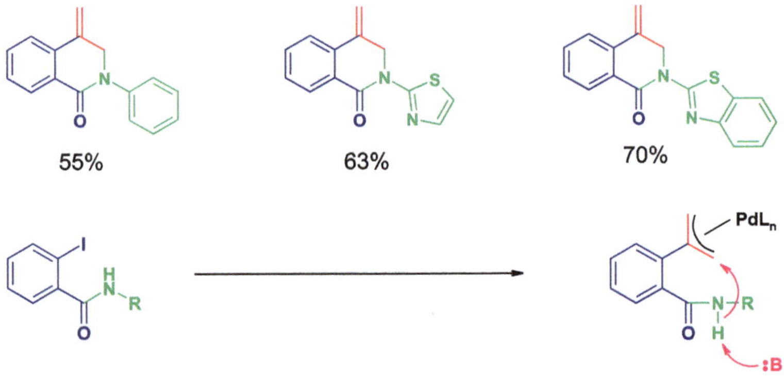

Fig. 8.

at 70–80°C. The room temperature generation of palladium nanoparticles from our PdCys is readily achieved in the presence of carbon monoxide. Hence carbonylation processes are especially suited to this type of catalyst. A wide range of three-component cascade processes have been devised using this approach together with carbon monoxide and allenes as substrates. The latter also significantly lowers the required reaction temperature

Acknowledgements. We thank Johnson Matthey, Avecia, GlaxoSmithkline, the EPRSC, and Leeds University for support.

References

Albisson DA,Bedford RB, Lawerence SE, Scully PN (1998) Orthopalladated triaryl phosphite complexes as highly active catalysts in biaryl coupling reactions. Chem Commun 2095–2096

Alonso DA, Najera C, Pacheco C (2002) Oxime-derived palladium complexes as very efficient catalysts for the Heck-Mizoroki reaction. Adv Synth Catal 344:172–183

Amatore C, Jutand A (1998) Coord Chem Rev 178–180:11–528

Amatore C, Jutand A (2000) Anionic Pd(0) and Pd(II) intermediates in palladium-catalyzed Heck and cross-coupling reactions. Acc Chem Res 33:314–321

Anwar U, Grigg R, Rasparini M, Savic V, Sridharan V (2000) Palladium catalysed cyclisation-Barbier-type allylation cascades Chem Commun 933–934

Bedford RB, Cazin CSJ, Hazelwood SL (2002a) Simple mixed tricyclohexylphosphane-triarylphosphite complexes as extremely high-activity catalysts for the Suzuki coupling of aryl chlorides. Angew Chem Int Ed 41:4120–4122

Bedford RB, Limmert ME, Hazelwood SL, Albisson DA (2002b) Platinum-catalysts for Suzuki biaryl coupling reactions. Organometallics 21:2599–2600

Beletskaya IP, Cheprakov AV (2000) The Heck reaction as a sharpening stone of palladium catalysis. Chem Rev 100:3009–3066

Cleghorn LAT, Cooper IR, Fishwick CWG, Grigg R, MacLachlan WS, Rasparini M, Sridharan V (2003a) Three-component bimetallic (Pd/In) mediated cascade allylation of C=X functionality: Part 1. Scope and class 1 examples with aldehydes and ketones. J Organomet Chem 687:483–493

Cleghorn LAT, Cooper IR, Grigg R, MacLachlan WS, Sridharan V (2003b) Additive effects in palladium–indium mediated Barbier type allylations. Tetrahedron Lett 44:7969–7973

Cleghorn LAT, Grigg R, MacLachlan WS, Sridharan V (2005) Bimetallic catalytic synthesis of annelated benzazepines. Chem Commun 3071–3073

Cooper IR, Grigg R, MacLachlan WS, Sridharan V, Thomas WA (2002a)-Diastereo- and regioselective palladium–indium bimetallic cyclisation–Barbier-type allylation cascades. Tetrahedron Lett 44:403–405

Cooper IR, Grigg R, MacLachlan WS, Thornton-Pett M, Sridharan V (2002b) 3-Component palladium–indium mediated diastereoselective cascade allylation of imines with allenes and aryl iodides. Chem Commun 1372–1373

Curtin ML, Frey RR, Heyman HR, Serris KA, Steenman DH, Holmes JH, Bousquet PF, Gunha GA, Moskey MD, Ahmed AA, Pease LJ, Glare KB, Stewart KD, Davidren SK, Michaelides M (2004) Isoindolinone ureas: a novel class of KDR kinase inhibitors. Bioorg Med Chem Lett 14:4505–4509

Dupont J, Consorti CS, Spencer J (2005) The potential of palladacycles: more than just precatalysts. Chem Rev 105:2527–2571

Evans P, Hogg P, Grigg R, Nurnabi M, Hinsley J, Sridharan V, Suganthan S, Korn S, Collard S, Muir JE (2005) 8-Methylquinoline palladacycles: stable and efficient catalysts for carbon–carbon bond formation. Tetrahedron 61:9696–9704

Funato N, Takayanagi H, Konda Y, Harigaya Y (1994) Absolute configuration of staurosporine by x-ray analysis. Tetrahedron Lett 35:1251–1254

Gai X, Grigg R, Collard S, Muir J (2000a) Palladium catalyzed intramolecular nucleophilic addition of allylic species, generated from allene, to aryl aldehydes and ketones. Chem Commun 1765–1766

Gai X, Grigg R, Collard S, Muir J (2000a) Palladium catalyzed intramolecular nucleophilic addition of allylic species, generated from allene, to aryl aldehydes and ketones. Chem Commun 1765–1766

Gai X, Grigg R, Ramzan MI, Sridharan V Collard S, Muir JE (2000b) Pyrazole and benzothiazole palladacycles: stable and efficient catalysts for carbon–carbon bond formation. Chem Commun 2053–2054

Gai X, Grigg R, Khamnaen T, Rajviroongit S, Sridharan V, Zhang L, Collard S, Keep A (2003a) Synthesis of 3-substituted isoindolin-1-ones via a palladium-catalyzed 3-component carbonylation/amination/Michael addition process. Tetrahedron Lett 44:7441–7443

Gai X, Grigg R, Koppen I, Marchbank J, Sridharan V (2003b) Synthesis of carbo- and heterocycles via a palladium-catalysed allene insertion–nucleophile incorporation–Michael addition cascade. Tetrahedron Lett 44:7445–7448

Gai X, Grigg R, Khamnaen T, Rajviroongit S, Sridharan V, Zhang L, Collard S, Keep A (2005) Synthesis of N-substituted isoindolinones via a palladium catalyzed three-component carbonylation–amination–Michael addition cascade. Can J Chem 83:990–1005

Green SJ, Swartz EM, Shah JH, Madsen J, D'Amato RJ (2001) USP 6071948

Grigg R, Khamnaen T, Rajviroongit S, Sridharan V (2002) Synthesis of *N*-substituted 4-methylene-3,4-dihydro-1(2*H*)-isoquinolin-1-ones via a palladium-catalysed three-component process. Tetrahedron Lett 43:2601–2603

Grigg R, Zhang L, Collard S, Keep A (2003) Isoindolinones via a room temperature palladium nanoparticle-catalysed 3-component cyclative carbonylation–amination cascade. Tetrahedron Lett 44:6979–6982

Grigg R, Zhang L, Collard S, Ellis P, Keep A (2004) Facile generation and morphology of Pd nanoparticles from palladacycles and carbon monoxide. J Organomet Chem 689:170–173

Gruber AS, Zim D, Ebeling G, Monteiro AL Dupont (2000) Sulfur-containing palladacycles as catalyst precursors for the Heck reaction. J Org Lett 2:1287–1290

Herrmann WA, Muchlhofer M (2002) In: Cornils B, Herman WA (eds) Applied homogenous catalysts with organometallic compounds, 2nd edn., Vol. 3. Wiley-VCH, Weinhein, Germany, pp 1086–1093

Herrmann WA, Brossmer C, Oefele K, Reisinger CP, Priermeie T, Beller M, Fisher H (1995) Coordination chemistry and mechanisms of metal-catalyzed C–C coupling reactions. Part 5. Palladacycles as structurally defined catalysts for the Heck olefination of chloro- and bromoarenes. Angew Chem Int Ed Engl 34:1844–1847

Herrmann WA, Brossmer C, Reisinger CP, Riermaier T, Ofele K, Beller M (1997) Coordination chemistry and mechanisms of metal-catalyzed C–C coupling reactions. Part 10. Palladacycles: efficient new catalysts for the Heck vinylation of aryl halides. Chem Eur J 3:1357–1364

Iyer S, Jayanthi A (2001) Acetylferrocenyloxime palladacycle-catalyzed Heck reactions. Tetrahedron Lett 42:7877–7878

Iyer S, Ramesh C (2000) Aryl–Pd covalently bonded palladacycles, novel amino and oxime catalysts {di-μ-chlorobis(benzaldehydeoxime-6-*C*,*N*)dipalladium(II), di-μ-chlorobis(dimethylbenzylamine-6-*C*,*N*)dipalladium(II)} for the Heck reaction. Tetrahedron Lett 41:8981–8984

Jeffery T (1984) Palladium-catalysed vinylation of organic halides under solid–liquid phase transfer conditions. J Chem Soc Chem Commun 1287–1289; (b) idem,

Jeffery T (1996) In: Liebeskind LS (ed) Advances in metal-organic chemistry. Vol. 5. JAI Press, London, pp 153–260

Kopple K, Meyerstein D, Meisel D (1980) Mechanism of the catalytic hydrogen production by gold sols. Hydrogen/deuterium isotope effect studies. J Phys Chem 84:870–875

Louie J, Hartwig JF (1996) A route to Pd0 from PdII metallacycles in amination and cross-coupling chemistry. Angew Chem Int Ed Engl 35:2359–2361

Mucalo MR, Coouey RP (1989) F.T.I.R. spectra of carbon monoxide adsorbed on platinum sols. J Chem Soc Chem Commun 94–95

Munoz MP, Martin-Matute B, Fernandez-Rivas C, Cardenas DJ, Echavarren AM (2001) Palladacycles as precatalysts in Heck and cross-coupling reactions. Adv Synth Catal 343:338–342

Norman MH, Minick DJ, Rigdon GC (1996) Effect of linking bridge modifications on the antipsychotic profile of some phthalimide and isoindolinone derivatives. J Med Chem 39:149–157

Ohff M, Ohff A, Milstein D (1999) Highly active Pd^{II} cyclometallated imine catalysts for the Heck reaction. Chem Commun 357–358

Palencia H, Garcia-Jimenez F, Takacs JM (2004) Suzuki–Miyaura coupling with high turnover number using an *N*-acyl-*N*-heterocyclic carbene palladacycle precursor. Tetrahedron Lett 45:3849–3853

Powell DR, Dahl LF (2000) Nanosized $Pd_{145}(CO)_x(PEt_3)_{30}$ Containing a capped three-shell 145-atom metal-core geometry of pseudo icosahedral symmetry. Angew Chem Int Ed 39:4121–4125

Rosner T, Le Bars J, Pfaltz A, Blackmond DG (2001a) Kinetic studies of Heck coupling reactions using palladacycle catalysts: experimental and kinetic modeling of the role of dimer species. J Am Chem Soc 123:1848–1855

Rosner T, Pfaltz A, Blackmond DG (2001b) Observation of unusual kinetics in Heck reactions of aryl halides: the role of non-steady-state catalyst concentration. Ibid 4621–4622

Roucoux A, Schultz J, Patin H (2002) Reduced transition metal colloids: a novel family of reusable catalysts. Chem Rev 102:3757–3778

Toyooka K, Kanamitsu N, Yoshimra M (2004) PCT international application WO2004/048332

Van Srijdonck GPF, Boete MDK, van Leeuwen PWN (1999) Fast palladium catalyzed arylation of alkenes using bulky monodentate phosphorus ligands. Eur J Inorg Chem 1073–1076

Weissman H, Milstein D (1999) Highly active Pd^{II} cyclometallated imine catalyst for the Suzuki reaction. Chem Commun 1901–1902

Ernst Schering Foundation Symposium Proceedings, Vol. 3, pp. 99–116
DOI 10.1007/2789_2007_030

Published Online: 23 May 2007

Applying Homogeneous Catalysis for the Synthesis of Pharmaceuticals

M. Beller(✉)

Leibniz-Institut für Katalyse e.V., University of Rostock, Albert-Einstein-Str. 29a, 18059 Rostock, Germany
email: *matthias.beller@catalysis.de*

Abstract. This article describes recent achievements of my research group in the Leibniz-Institut für Katalyse e.V. in the area of applied homogeneous catalysis for the synthesis of biologically active compounds. Special focus is given on the development of novel and practical palladium and copper catalysts for the functionalization of haloarenes and haloheteroarenes.

1 Introduction

Catalysis is known as the science of accelerating chemical transformations. In general, various starting materials are converted to more complex molecules with versatile applications. Traditionally, catalysts are divided into homogeneous and heterogeneous catalysts, biocatalysts (enzymes), photocatalysts, and electrocatalysts, which are mainly used in environmental protection, production of chemicals, oil processing in refineries, and polymer synthesis. With respect to life sciences, it is important to note that most currently marketed drugs as well as new biologically active compounds come into contact with catalysts over the course of their synthesis. Catalysts ensure that reactions proceed with high efficiency, high yield, and avoid unwanted by-products. Importantly, they often allow for a more economical production compared to classical stoichiometric procedures. Thereby catalysis enables the chemical, agrochemical, and pharmaceutical industry to offer a wide range of products for our health, environment, and nutrition (Fig. 1). Obviously, catalysts are indispensable for the needs of today's society and an important tool for increasing sustainability.

Comparing the different subareas of catalysis, all have common characteristics, but also significant differences are visible. Our area of main interest is homogeneous catalysis based on transition metal catalysts. In general, these homogeneous catalysts (most often the precatalyst) are molecularly defined.

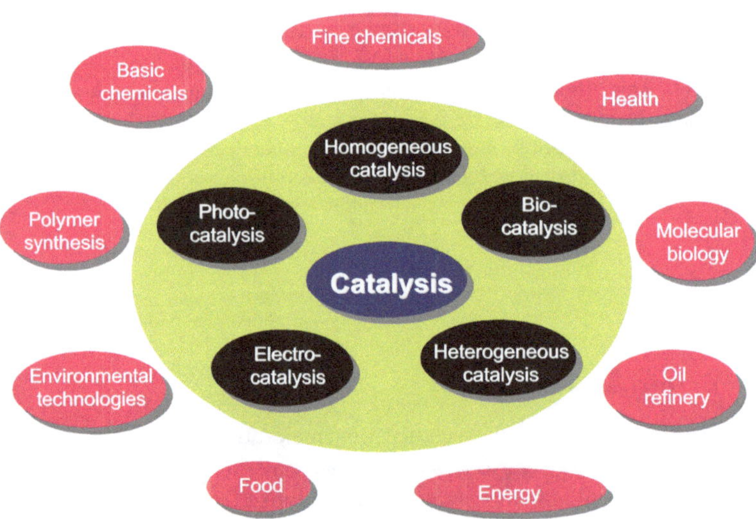

Fig. 1. Different areas of catalysis and their applications

This fact combined with the possibility to synthesize stable intermediates of the catalytic cycle allows an easier understanding of the reaction mechanism of a given catalytic process. Thus, improved catalysts are more rational to develop compared to other areas of catalysis. For the synthesis of biologically active molecules, it is important that homogeneous catalysis takes place under comparably mild reaction conditions. Here, typically reactions proceed between room temperature and 120°C in liquid phase in a batch-wise mode. Thus, it is not surprising that the conversion of more complicated organic building blocks with various functional groups is predominantly performed with homogeneous catalysts. Nevertheless, a recent survey in a pharmaceutical company showed that only 5%–10% of the steps in drug production are catalyzed by homogeneous catalysts (Carey et al. 2006). Clearly, there is significant room for improvement. For the development of new and better catalysts, it is crucial to tune the activity, productivity, and selectivity of a given metal center or in other words to control the electronic and steric properties of each active center. Therefore, key issues are the synthesis of molecularly defined catalytic centers and organic ligands, which control the catalyst center. Thus, ligand-tailoring constitutes an extremely powerful tool to control all kinds of selectivity in a given catalytic reaction and to influence catalyst stability and activity. Due to the advancements in organometallic chemistry and organic ligand synthesis, nowadays a plethora of ligands [P-, N-, and recently C-ligands) is theoretically available (>100,000). A selection of well-established ligands for homogeneous catalysis is shown in Scheme 1.

Apart from the classic aryl and alkyl phosphines and amines, recently carbenes have become more important. In addition, mixed ligand systems with two different chelating groups, as well as hemilabile ligands, find increasing interest. Despite all progress in ligand synthesis, the bottleneck for high-throughput testing of novel homogeneous catalysts is often the availability of useful ligand libraries.

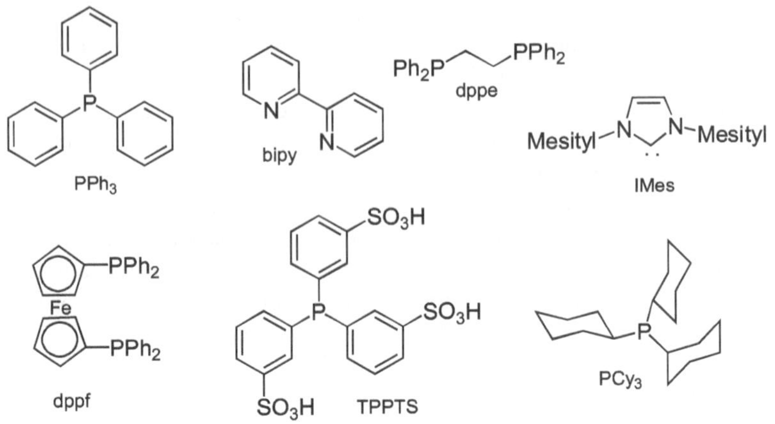

Scheme 1. A selection of known ligands for homogeneous catalysis

2 Research Interest

In general, we use homogeneous catalysis and organometallic chemistry for a practical organic synthesis of fine chemicals, biologically active compounds, and new materials. More specifically, the current work of my research group can be divided into six subareas (Fig. 2). Since the early 1990s, the development of synthetic methods and the application of catalysis for the preparation of biologically active compounds has been a major interest. In this respect, especially palladium catalysis (Zapf and Beller 2004) and amination reactions (Seayad et al. 2002; Beller et al. 2002) have been investigated over the years.

After joining the Leibniz-Institut für Katalyse e.V. an der Universität Rostock (LIKAT) in 1998, my group began investigating the prominent areas of catalysis with carbon monoxide (Moballigh et al. 2003; Klein et al. 2005) and oxidation catalysis. Most recently, we also initiated a program on the use of homogeneous catalysis for hydrogen generation and biomimetic catalysis. In addition to basic science, an important goal of my group as well as the LIKAT is the facilitation of innovation processes by transfer of basic research to industrially useful purposes.

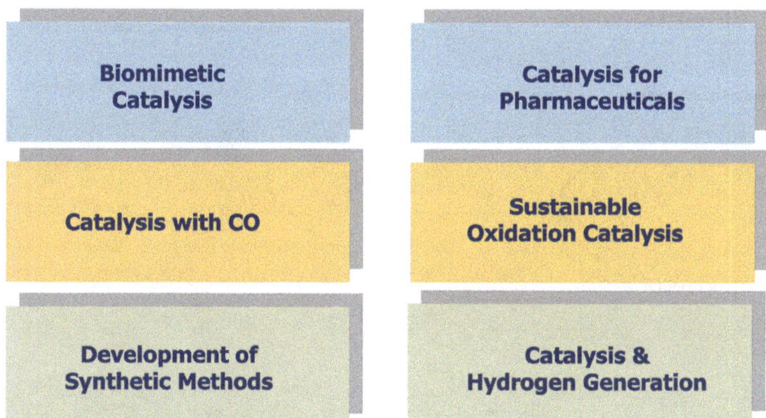

Fig. 2. Research areas of interest

In the last 3 years, this concept has been successful and several catalysts developed at the Leibniz-Institut für Katalyse have been transferred to industry. Among these, a palladium catalyst is nowadays used for the synthesis of pharmaintermediates on the several-thousand kilogram scale.

3 The Development of Efficient Catalysts for the Functionalization of Haloarenes

In the past few years, pharmaceutical products have been produced using predominantly stoichiometric organic synthesis. However, since the 1990s homogeneous catalysis is being used more and more often for the production of current drugs. Among the many different catalytic transformations, palladium-catalyzed C–C and C–N bond forming reactions received special attention and consequently had a particularly high impact on organic synthesis (Beller and Bolm 2004). The increasing popularity of these methods in synthesis is seen in each issue of modern scientific journals dedicated to organic synthesis, organometallic chemistry, and catalysis. The main advantages of the coupling processes shown in Scheme 2 are based on the ready availability of starting

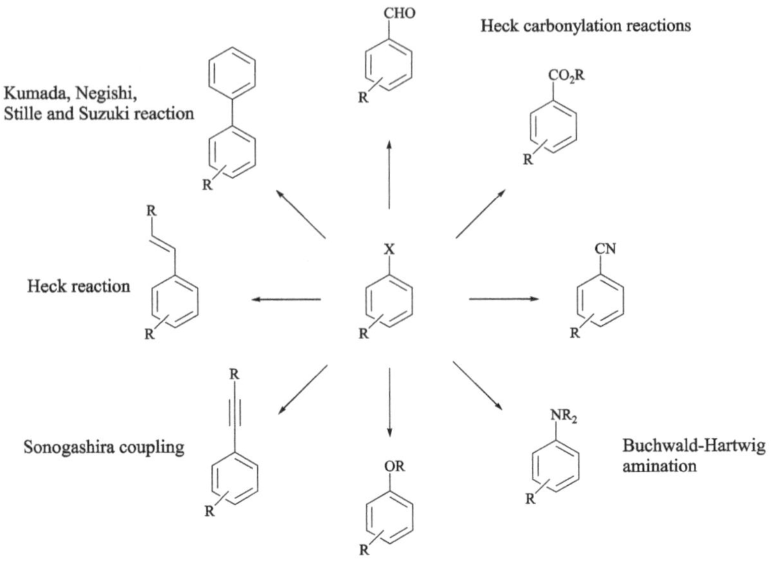

CHO
Heck carbonylation reactions
Kumada, Negishi, Stille and Suzuki reaction
CO_2R
Heck reaction
X
CN
Sonogashira coupling
NR_2
Buchwald-Hartwig amination
OR

Scheme 2. Selected examples of palladium-catalyzed coupling reactions

materials, the simplicity and generality of the methods, as well as the broad tolerance of palladium catalysts toward various functional groups.

With regard to application in the pharmaceutical industry, palladium-catalyzed coupling reactions offer the opportunity of shorter and more selective routes for a number of currently marketed and future drugs. Therefore, it is not surprising that since the early 1990s more and more palladium-catalyzed reactions are transferred from academic protocols to the industrial context (Beller et al. 2001; Beller and Zapf 2002; de Vries 2001). Selected examples of processes that are used nowadays or have been used in the pharmaceutical industry are shown in Scheme 3. In order to see more realizations of this type of chemistry, more active and productive palladium catalysts have to be developed because of the high price of palladium and most often the ligand system.

With respect to new ligands, recently we have investigated the scope and limitations of di(1-adamantyl)alkylphosphines and *N*-aryl-2-(dialkylphosphino)pyrrole. Because of their facile synthesis and their ex-

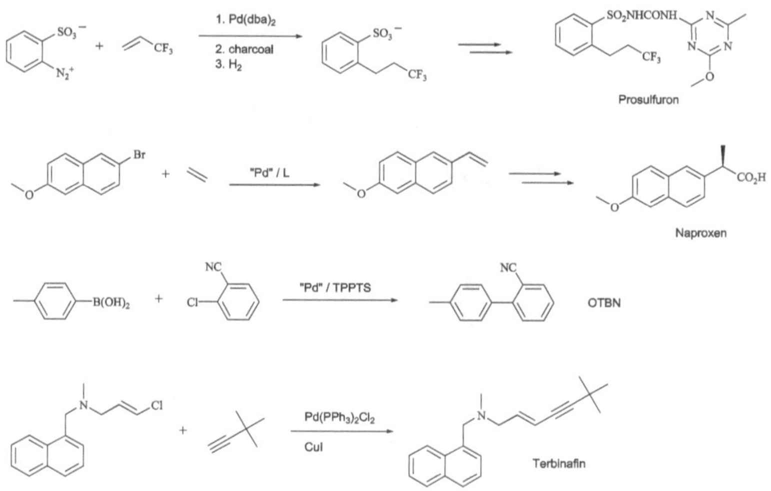

Scheme 3. Selected palladium-catalyzed industrial processes for the synthesis of pharmaintermediates

cellent performance, technical processes based on these ligands have already been launched (see below). Initially, it was Andreas Ehrentraut, a PhD student who in 1999 synthesized so-called adamantylphosphines, a class of ligands that had been scarcely known until that time and that had never been applied in palladium-catalyzed reactions before our work.

From 2000 on, this class of ligands has been developed in joint cooperation with Degussa Homogeneous Catalysts (DHC). Together, we were pleased to find that a ligand library of alkyladamantylphosphines could be easily synthesized in three steps from easily available adamantane. Among the various alkyladamantylphosphines, di(1-adamantyl)-*n*-butylphosphine (**1**, cata*CX*ium® A) is an excellent ligand for different functionalization reactions of aryl and heteroaryl bromides. Even deactivated aryl chlorides can be used efficiently in palladium-catalyzed coupling reactions. Nowadays, gram quantities of cata*CX*ium® A and similar derivatives are available from Strem (Fig. 3). Quantities up to 100 kg are available from Degussa Homogeneous Catalysts.

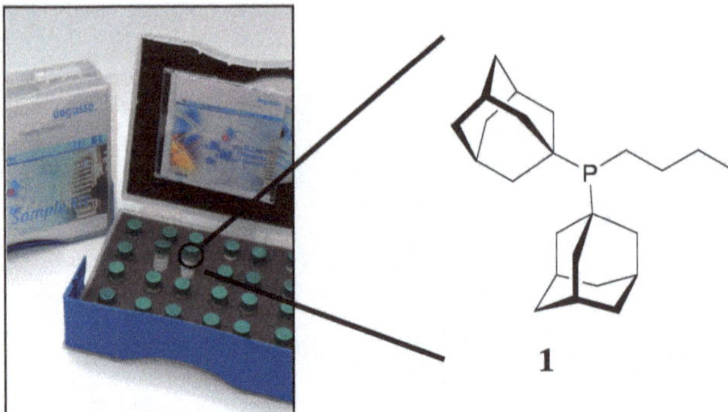

Fig. 3. Ligand toolbox with di(1-adamantyl)-*n*-butylphosphine (**1**, cata*CX*ium® A)

An important advantage of the adamantylphosphines compared to similar well-known *tert*-butylphosphines is their significantly increased stability toward air and moisture. In general, **1** can be handled for a short time on air without problems. This makes the ligand and resulting catalysts especially attractive for larger-scale applications. The new ligands have been applied successfully by us and other research groups in Heck (Ehrentraut et al. 2000), Suzuki (Zapf et al. 2000), and Buchwald-Hartwig amination (Ehrentraut et al. 2002) reactions of different aryl halides. They are also useful for the α-arylation of ketones and Sonogashira reactions (Ehrentraut et al. 2002).

Selected coupling reactions of aryl chlorides in the presence of Pd/1 are shown in Scheme 4. For example, nonactivated and sterically hindered 2,6-dimethylchlorobenzene reacts with bulky, sterically congested anilines smoothly at low catalyst loading (0.5 mol% $Pd(OAc)_2$/**1**; P/Pd = 2:1).

More recently, we have been engaged in mechanistically related carbonylations of aryl halides (Klaus et al. 2006). As demonstrated in Scheme 5, such reactions offer numerous possibilities for the selective synthesis of aromatic carbonyl compounds.

To explore the opportunities of catalysis for drug development, we started a program on the synthesis of potentially active amphetamine

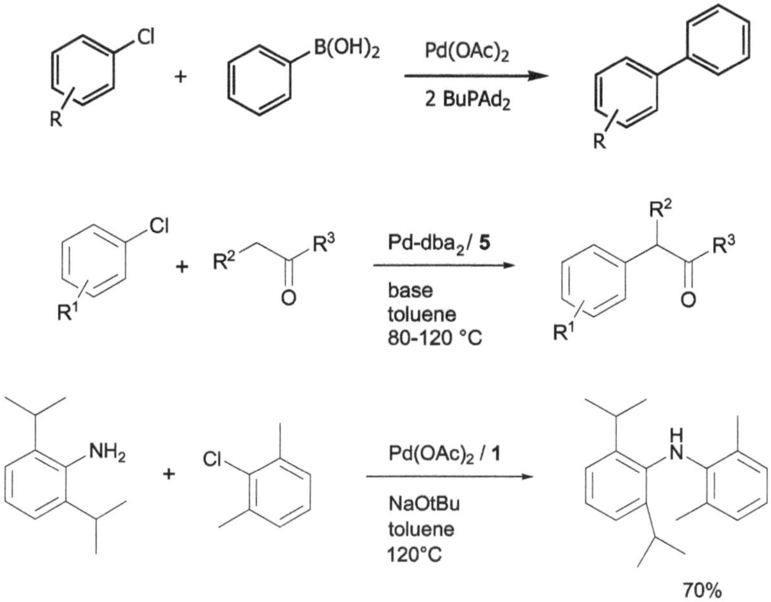

Scheme 4. Palladium-catalyzed coupling reactions of aryl chlorides

analogs using palladium-catalyzed carbonylations as a tool box (Kumar et al. 2004). As target we envisioned the serotonin (5-HT)-receptor subtype 2A receptor, for which phenethyl piperazines can be ligands. Different derivatives of this class of compounds are suitable for the treatment of a wide variety of diseases such as psychosis, schizophrenia, depression, neural disorders, memory disorders, Parkinson's disease, amyotrophic lateral sclerosis, Alzheimer's disease, Huntington's disease, eating disorders such as nervous bulimia and anorexia, premenstrual syndromes, and for positively influencing compulsive behaviors (obsessive-compulsive disorder, OCDA). We demonstrated that in only three catalytic steps 5-HT-2A receptor antagonists can be synthesized efficiently with high diversity. As an initial step, an anti-Markovnikov addition of amines to styrenes provides easy access to *N*-(arylalkyl)piperazines, which constitutes the core structure of the active molecules. After palladium-catalyzed debenzylation, the free amines were success-

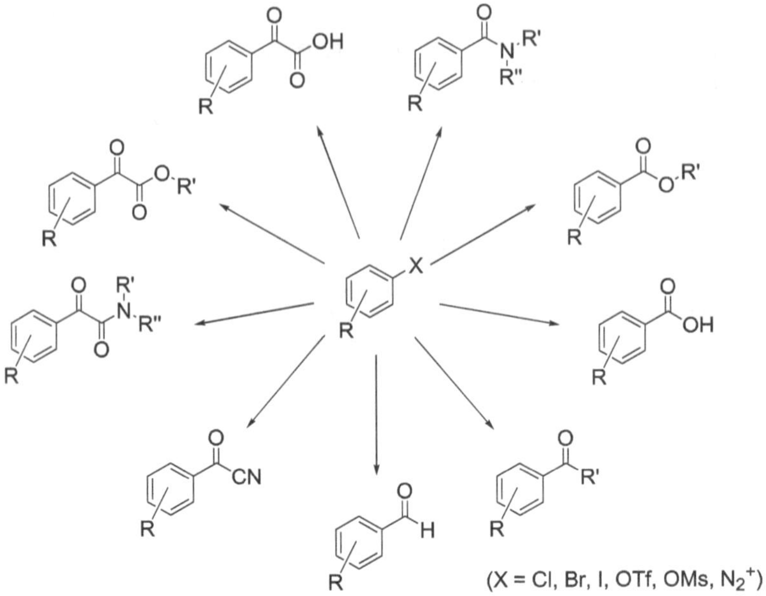

(X = Cl, Br, I, OTf, OMs, N_2^+)

Scheme 5. Palladium-catalyzed carbonylations of aryl halides and pseudohalides

fully carbonylated with different aromatic and heteroaromatic halides and carbon monoxide to yield the desired compounds in good to excellent yield. The key reaction, i.e., palladium-catalyzed aminocarbonylation of haloarenes/heterocycles, showed tolerance against various functional groups, thereby demonstrating that it was possible to synthesize a wide variety of new derivatives of this promising class of pharmaceuticals (Scheme 6.)

Among the different aromatic carbonyl compounds, aldehydes are probably the most useful class of products, as the highly reactive aldehyde group can be easily employed in numerous C–C-, and C–N-coupling reactions, reductions as well as other transformations.

Based on the use of di-1-adamantyl-*n*-butylphosphine (cata*CX*ium® A) **1** as ligand, a general and efficient palladium-catalyzed formylation of aryl and heteroaryl bromides has been developed by Degussa AG

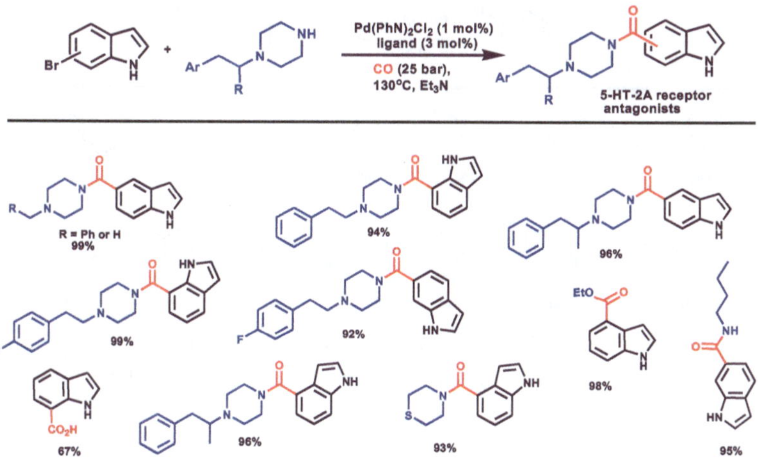

Scheme 6. Synthesis of 5-HT-2A receptor antagonists by catalytic carbonylations of haloindoles

and our group (Scheme 7) (Klaus et al. 2006). Advantageously, synthesis gas can be used as an environmentally benign formylation source at comparably low pressure (≤5 bar). Because of the high air stability of cata*CX*ium® A ligand, it is possible to load the autoclave with all the reagents at the air and then to purge it with synthesis gas without decreasing the catalyst activity.

In general, a broad range of substituted aromatic and heteroaromatic aldehydes are obtained at unprecedented low catalyst concentrations in excellent yield (up to 99%). The simplicity of the reaction conditions and the practicability and usefulness of this novel method allows for the first time to obtain such reactions on industrial scale.

Similarly, a general palladium-catalyzed alkoxycarbonylation of aryl and heteroaryl bromides has been developed in the presence of bulky monodentate phosphines (Neumann et al. 2006). Studies of the butoxycarbonylation of three model substrates again revealed the advantages of di-1-adamantyl-*n*-butylphosphine (cata*CX*ium® A) compared to other ligands. As shown in Scheme 8, this catalyst system provided the corresponding benzoic acid derivatives (ester, amides, acids) in high yield at low catalyst loadings (0.5 mol% Pd or below).

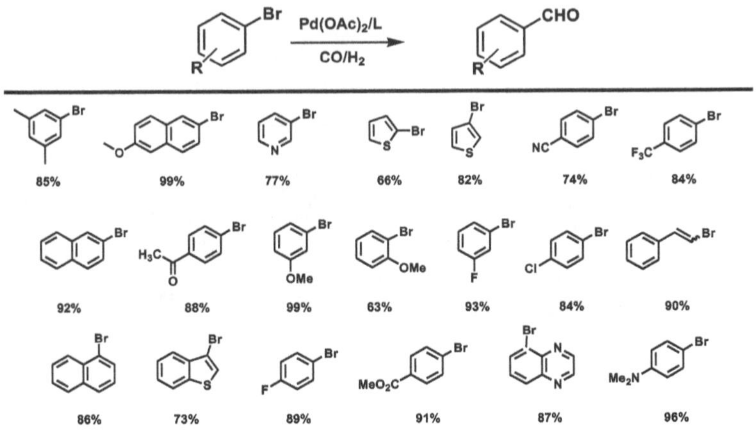

Scheme 7. Scope of the palladium-catalyzed formylation

Benzonitriles constitute another class of carboxylic acid derivatives. They are of considerable interest for organic synthesis as an integral part of herbicides, agrochemicals, pharmaceuticals, and natural products (Scheme 9) (Larock 1989). In general, the introduction of a cyanide group is the most direct and versatile route to prepare functionalized aryl nitriles. Moreover, this straightforward C1-conversion is an important elementary step for the synthesis of various carboxylic acid derivatives and the nitrile group has been shown to serve as a precursor for various functional groups: benzylamines, benzaldehydes, heterocycles, etc.

For more than a century, stoichiometric methods were presumed in the preparation of benzonitriles in laboratory and industry. These particularly include the Rosenmund-von Braun reaction of aryl halides, the diazotization of anilines and subsequent Sandmeyer reaction, and the ammoxidation. Because of (over)stoichiometric amounts of metal waste, lack of functional group tolerance, and harsh reaction conditions, these methods do not meet the criteria of modern sustainable synthesis.

In the last three decades, the development of transition metal catalyzed CC-coupling processes has revolutionized aromatic functional-

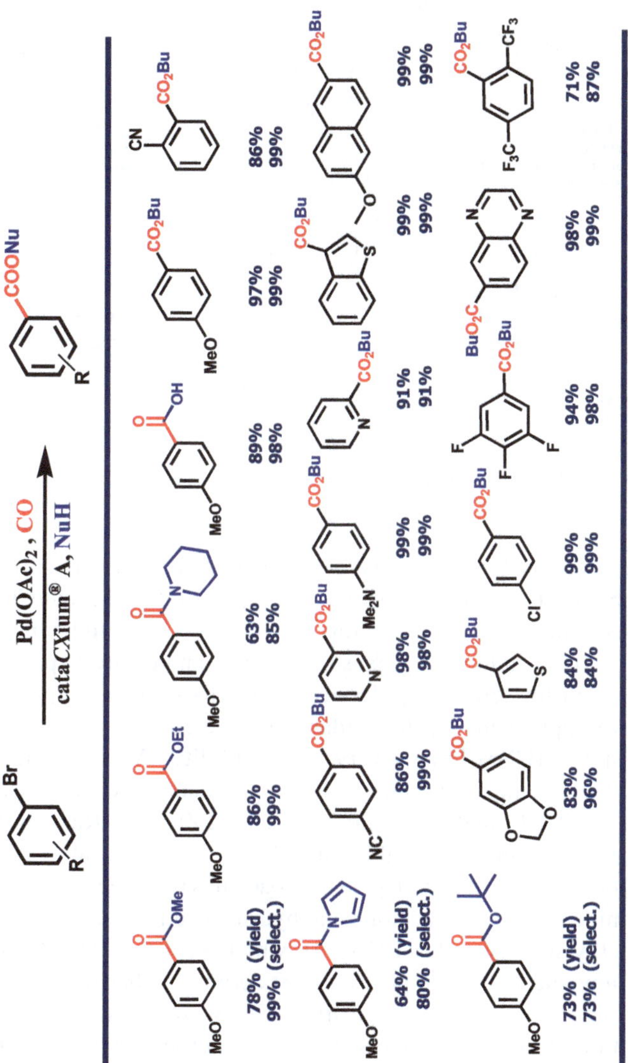

Scheme 8. Palladium-catalyzed carboxylations of aryl and heteroaryl bromides

Bicalutamid
Casodex (Zeneca)
non-steroidal antiandrogen, antineoplastic, anti(prostate)cancer

Periciazine
Aolept (Bayer)
antipsychotic, neuroleptic

Cyamemazine
Neutromil (Farmitalia)
neuroleptic, tranquilizer

Fadrozole
Arensin (Ciba-Geigy)
antineoplastic, non-steroidal aromatase inhibitor

Letrozole
Femara (Novartis Pharma)
antineoplastic, aromatase inhibitor

Citalopram
Cipramil (Promonta Lundbeck)
antidepressant, selective serotonin-uptake inhibitor

Scheme 9. Selected examples of biologically active benzonitriles

ization reactions. Unfortunately, metal-catalyzed cyanations suffer from the high affinity of cyanide toward the typical Pd-, Ni-, and Cu-based catalysts. Thus, a fast deactivation of the catalytic system by the formation of stable cyanide complexes is observed and catalysis proceeds in general with low efficiency. To overcome this problem, typically solvents are applied in which standard cyanide sources such as NaCN, KCN, and $Zn(CN)_2$ have a very low solubility. We and others have shown that organic, e.g., tetramethylethylendiamine (tmeda) (Sundermeier et al. 2001), and inorganic, e.g., Zn and Zn salts (Ramnauth et al. 2003), additives are beneficial for the reformation of the catalytically active metal center. Another elegant approach has been the slow dosage of the cyanide source, e.g. acetoncyanohydrine (Sundermeier et al. 2003) or trimethylsilyl cyanide (TMSCN), which kept the cyanide concentration low and led to a higher catalyst activity. Other recent developments included microwave activation (Chobanian et al. 2006) and the application of new catalyst systems (Jensen et al. 2005; Chidambaran 2004; Veauthier et al. 2005). However, all these developments still have drawbacks such as toxicity of the cyanide source and comparably high catalyst costs. In 2004, we described for the first time catalytic cyana-

Pd(OAc)2,
dppf,
X
R
CN
T. Schareina, A. Zapf, M. Beller, Chem. Commun. 2004, 1388-1389.
0.01mol% Pd, 120°C
0.25 mol% Pd, 160°C Yield = 75% TON = 300
0.1 mol% Pd, 120°C Yield = 99% TON = 990
0.1 mol% Pd, 140°C Yield = 86% TON = 860
0.1mol% Pd, 120°C Yield = 51% TON = 510
0.1 mol% Pd, 120°C Yield = 95% TON = 950
0.1 mol% Pd, 140°C Yield = 87% TON = 870
0.1 mol% Pd, 140°C Yield = 78% TON = 780
0.1 mol% Pd, 120°C Yield = 94% TON = 940

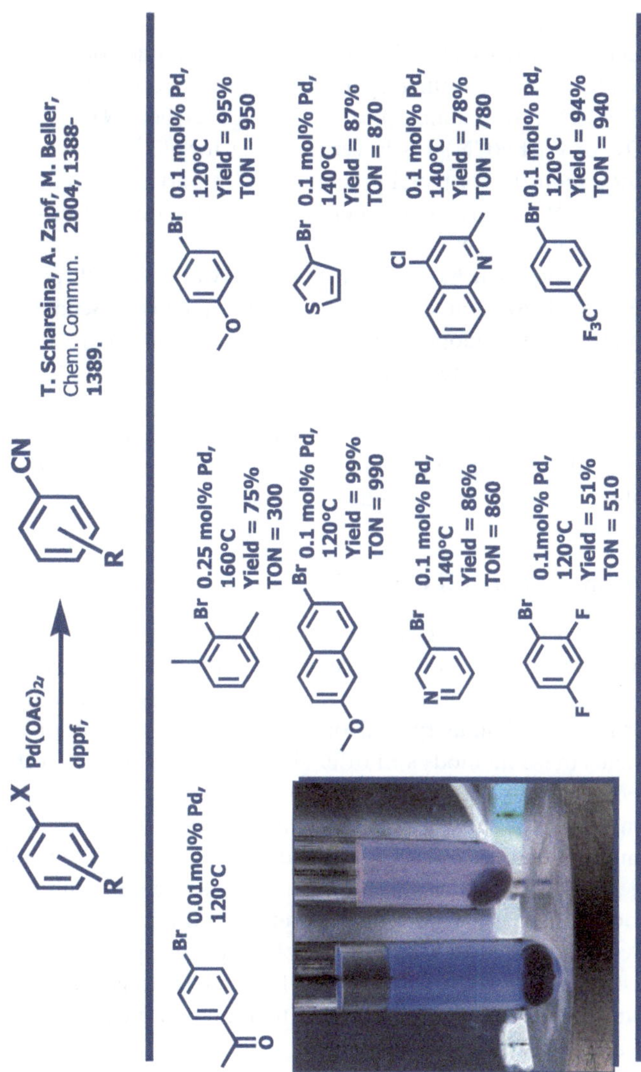

Fig. 4.

tions with potassium hexacyanoferrate(II) $K_4[Fe(CN)_6]$, which has the advantage of being essentially the least toxic cyanide source conceivable. While all known other cyanation sources are highly poisonous, e.g., KCN LD_{Lo} (oral, human) = 2.86 mg/kg, $K_4[Fe(CN)_6]$ is nontoxic (the LD_{50} of $K_4[Fe(CN)_6]$ is lower than that for NaCl!) and even used in the food industry for metal precipitation. Important for practical applications, $K_4[Fe(CN)_6]$ is commercially available on the ton-scale and even cheaper than KCN.

Our new approach has proven its initial value in both palladium- (Schareina et al. 2004) and copper-catalyzed cyanations (Schareina et al. 2005) and has been adopted by other groups. Very recently, in a joint collaboration with Saltigo GmbH we developed a new and improved copper-based catalyst system, which allows for efficient cyanations of a variety of aromatic and heteroaromatic halides. Importantly, notoriously difficult substrates react in excellent yield and selectivity, making the method applicable on an industrial scale.

4 Conclusion and Outlook

Nowadays, cross-coupling reactions allow for reliable CC-, CN-, and CO-bond formation in the synthesis of all kinds of biologically active compounds. A number of reactions are even efficient enough to run in the pharmaceutical industry on a ton scale. Nevertheless, despite all advancements these methods still offer significant challenges for the next years. For example, catalyst efficiency (activity and productivity) needs to be improved for substrates such as (nitrogen)heteroarenes and more functionalized coupling partners. In addition, another challenge for the coming years will be the development of cross-coupling alternatives, which do not use aryl halides as substrates but directly employ arenes. The advantages of such methods are obvious: cheaper substrates and less waste. Here, the development of selective CH-transformation reactions, either via metal-catalyzed CH-activation or via catalytic Friedel-Crafts variants, is the crucial point.

Acknowledgements. The work on the development of transition metal catalysts described herein was funded by Degussa AG, Saltigo GmbH, Merck KGA, the BMBF, the FCI, the Alexander von Humboldt-Stiftung, the Max-Buchner-

Forschungsstiftung, and the state of Mecklenburg-Vorpommern. I am thankful to all members of my group who worked for nearly a decade on the development of new catalysts for the synthesis of biologically active compounds, especially Dr. Alexander Zapf, Dipl. Chem. Andreas Ehrentraut, Dr. Anja Frisch, Dr. Mario Gomez, Dr. Surendra Harkal, Dr. Ralf Jackstell, Dr. Jürgen Krauter, Dr. Kamal Kumar, Dr. Wolfgang Mägerlein, Dr. Thomas Schareina, and Dr. Kumaravel Selvakumar. In addition, the work presented here would have not been possible without trustful and active industrial cooperation partners in this area. Here, I am very thankful to Dr. Thomas Riermeier, Dr. Axel Monsees, Dr. Juan Almena, Renat Kadyrov (all Degussa AG, Dr. Wolfgang Mägerlein, Dr. Nikolaus Müller (Saltigo GmbH), and Dr. Michael Arlt (Merck KGA).

References

Beller M, Bolm C (eds) (2004) Transition metals for organic synthesis, 2nd edn. Wiley-VCH, Weinheim

Beller M, Zapf A (2002) In: Negishi EI (ed) Organopalladium chemistry for organic synthesis, Vol. 1, Wiley-Interscience, New York, pp 1209–1222

Beller M, Zapf A, Mägerlein W (2001) Chem Eng Techn 24:575–582

Beller M, Breindl M, Eichberger M, Hartung CG, Seayad J, Thiel O, Tillack A, Trauthwein H (2002) Synlett, p 1579–1594

Carey JS, Laffan D, Thomson C, Williams MT (2006) Analysis of the reactions used for the preparation of drug candidate molecules. Org Biomol Chem 4:2337–2347

Chidambaram R (2004) Tetrahedron Lett 45:1441–1444

Chobanian HR Fors BP, Lin LS (2006) Tetrahedron Lett 47:3303–3306

de Vries JG (2001) Can J Chem 79:1086–1092

Ehrentraut A, Zapf A, Beller M (2000) Synlett, pp 1589–1592

Ehrentraut A, Zapf A, Beller M (2002a) J Mol Cat 182–183:515–523

Ehrentraut A, Zapf A, Beller M (2002b) Adv Synth Catal 344:209–217

Jensen RS, Gajare AS, Toyota K, Yoshifuji M, Ozawa F (2005) Tetrahedron Lett 46:8645–8647

Klaus S, Neumann H, Zapf A, Strübing D, Hübner S, Almena J, Riermeier T, Groß P, Sarich M, Krahnert W-R, Rossen K, Beller M (2006) A general and efficient method for the formylation of aryl and heteroaryl bromides. Angew Chem Int Ed 45:154–156

Klein H, Jackstell R, Beller M (2005) Synthesis of linear aldehydes from internal olefins in water. Chem Commun 2283–2284

Kumar K, Michalik D, Garcia Castro I, Tillack A, Zapf A, Arlt M, Heinrich T, Böttcher H, Beller M (2004) Biologically active compounds through catalysis: efficient synthesis of N-(heteroarylcarbonyl)-N'-(arylalkyl)piperazines. Chem Eur J 10:746–757

Larock RC (1989) Comprehensive organic transformations. VCH, New York

Moballigh A, Seayad A, Jackstell R, Beller M (2003) Amines made easily: a highly selective hydroaminomethylation of olefins. J Am Chem Soc 125:10311–10318

Neumann H, Brennführer A, Groß P, Riermeier T, Almena J, Beller M (2006) Adv Synth Catal 348:1255–1261

Ramnauth J, Bhardwaj N, Renton P, Rakhit S, Maddaford SP (2003) Synlett 2237–2238

Schareina T, Zapf A, Beller M (2004a) Potassium hexacyanoferrate(II)—a new cyanating agent for the palladium-catalyzed cyanation of aryl halides. Chem Commun 1388–1389

Schareina T, Zapf A, Beller M (2004b) J Organomet Chem 689:4576–4583

Schareina T, Zapf A, Beller M (2005) Tetrahedron Lett 46:2585–2588

Seayad J, Tillack A, Hartung CJ, Beller M (2002) Adv Synth Catal 344:795–813

Sundermeier M, Zapf A, Beller M (2001) J Sans Tetrahedron Lett 42:6707–6710

Sundermeier M, Zapf A, Beller M (2003) A convenient procedure for the palladium-catalyzed cyanation of aryl halides. Angew Chem Int Ed 42:1661–1664

Veauthier JM, Carlson CN Collis GE, Kiplinger JL, John KD (2005) Synthesis 2686–2683

Zapf A, Beller M (2005) The development of efficient catalysts for palladium-catalyzed coupling reactions of aryl halides. Chem Commun 431–440

Zapf A, Ehrentraut A, Beller M (2000a) Angew Chem 112:4315–4317

Zapf A, Ehrentraut A, Beller M (2000b) A new highly efficient catalyst system for the coupling of nonactivated and deactivated aryl chlorides with arylboronic acids palladium-catalyzed reactions for fine chemical synthesis, Part 17. Angew Chem Int Ed 39:4153–4156

Ernst Schering Foundation Symposium Proceedings, Vol. 3, pp. 117–131
DOI 10.1007/2789_2007_031

Published Online: 23 May 2007

An Integrated Approach to Developing Chemoenzymatic Processes at the Industrial Scale

J. Tao(✉), L. Zhao

BioVerdant, Inc., 7330 Carroll Road, 92121 San Diego, USA
email: *alex.tao@BioVerdant.com*

Abstract. Biocatalysis is becoming a transformational technology for chemical synthesis as a result of recent advances in enzyme discovery, structural biology, protein expression, high-throughput screening, and enzyme evolution technologies. To truly impact chemical synthesis at the industrial scale, biotransformations and chemical research and development must be integrated to develop cost-effective and environmentally friendly solutions for drug manufacture. In this chapter, some recent applications of chemoenzymatic synthesis of pharmaceutically active ingredients or advanced intermediates were selected to illustrate the principle of this integrated approach.

1 Contents

Biocatalysis is an emerging and transformational technology uniquely suited to the manufacture of active ingredients in the pharmaceutical and related industries as a result of recent breakthroughs in biotechnology: exponential growth in publicly available sequences from the gene database and protein data bank (PDB), efficient molecular cloning and protein expression platforms, and powerful, directed enzyme evolution technologies to improve a biocatalyst's specificity, selectivity, and stability. However, utilization of biotransformations at commercial scale is still limited, and most current applications for pharmaceutical synthesis have been to drug molecules at early stages before proof of concept (POC) (Leresche and Meyer 2006; Straathof et al. 2002). In order to develop cost-effective and environmentally friendly chemical synthesis at the industrial scale, it is essential to integrate biocatalysis and modern chemical research and development synergistically to deliver manufacturing routes with fewer synthetic steps, reduced waste streams, and improved overall synthetic efficiency including yields, regio- and stereoselectivity (Yazbeck et al. 2004).

As more and more enzymes become available, it is essential to develop an automated microtiter-based screening protocol, which allows the rapid identification of desired biocatalyst hits from a family of enzyme libraries using a minimal amount of substrates and enzymes. As a result of comprehensive screening, the success rate can be significantly improved and many unique enzymes or conditions can be identified, which were previously largely ignored in the synthetic community (Yazbeck et al. 2003).

For example, in the synthesis of AG7404, a rhinovirus protease inhibitor for the treatment of the common cold, the key intermediate is an acid precursor (Scheme 1). The existing chemical resolution is inefficient, suffering low yields. Through a 96-well plate-based screening of a comprehensive library of hydrolases, *Bacillus lentus* protease (BLP) was identified as the best hit. Priori to this work, the synthetic applications of BLP in process development have never been reported in the literature. In the presence of 35% acetone and 100 g/l of the racemic ester, excellent enantioselectivity (96% ee) was obtained at a conversion of 50% at pH 8.2 after 24 h (Scheme 1). Moreover, the wrong enantiomer can be readily recycled using a catalytic amount of DBU. Overall, the chemoenzymatic process is much more cost-effective and high-yielding than the classic resolution route (Martinez et al. 2004).

It is well known that solvents have a profound effect on protein conformation. It is often beneficial to apply the phenomena to the optimization of enzymatic reactions. For example, in the synthesis of the GABA

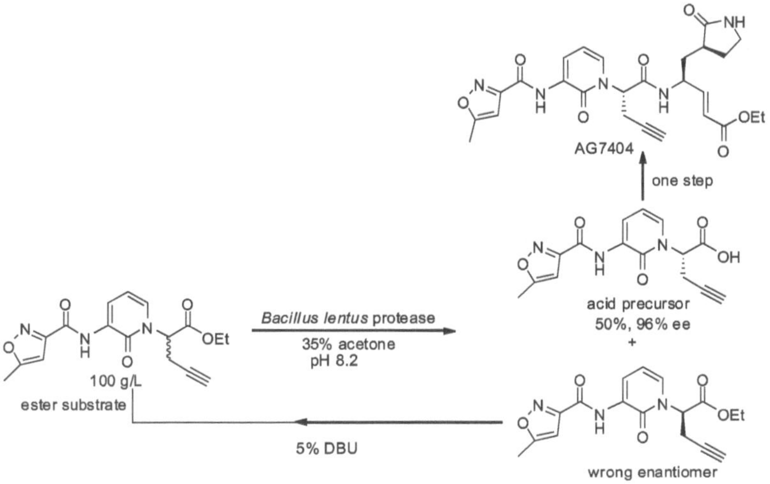

Scheme 1. Chemoenzymatic synthesis of a Rhinovirus protease inhibitor AG7404

(γ-amino butyric acid) inhibitor for the treatment of neuropathic pain and epilepsy, one ideal scenario is to identify an enzyme that can be diastereomerically indiscreet but enantiomerically selective since both the *cis*- and *trans*-diastereomer of the desired enantiomer can be converted into the final product in the subsequent steps (Scheme 2) (Yazbeck et al. 2006).

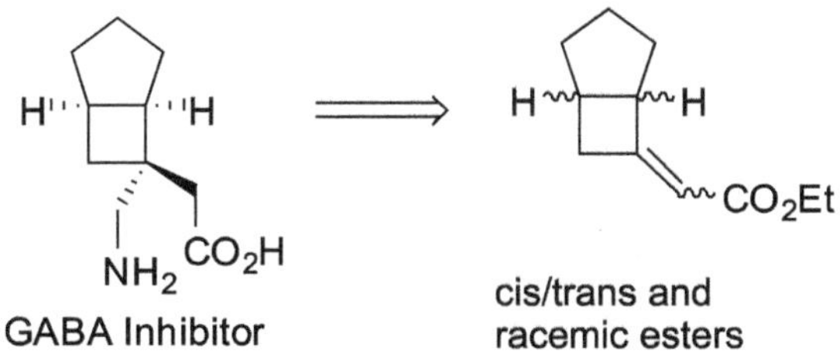

Scheme 2. Retrosynthetic analysis of the GABA inhibitor

Initial screening of a large library of enzymes gave only one hit, *Candida antarctica* lipase B, which has an E-value of 10, making it essentially useless for process development. Through extensive solvent screening and optimization, in the presence of 40% acetone, an E-value of more than 100 was obtained for the reaction and the chemoenzymatic route was quickly scaled up to prepare the GABA inhibitor for clinical trials (Scheme 3). Preliminary 2D-TROSY NMR study using N^{15}-labeled *Candida antarctica* lipase B indicated the dramatic solvent effect was probably caused by a global conformational change of the protein (Yazbeck et al. 2006).

One of the most important aspects of the integration of biocatalysis and chemical research and development is to apply both the chemical and biocatalytic retrosynthetic analyses at the very beginning of route scouting. After all, it is often difficult to fix a poor reaction sequence by replacing one problematic chemical step with an enzymatic transformation. For example, the first generation synthesis of Ruprintrivir™, a rhinovirus inhibitor was lengthy, with low overall yields and inefficient chromatographic separation (Dragovich et al. 1998). An uncommon bond disconnection for the ketomethylene peptidomimetic moiety (bracketed) resulted in a new construction of this complex drug molecule from a key building block P_2, which can be prepared by enzymatic reduction using lactate dehydrogenase (Scheme 4). It should be noted that chemical syntheses of this deceptively simple molecule have not been successful by either asymmetric hydrogenation or chiral pool approaches (Tao and McGee 2002). However, P_2 was obtained in high yields (80%–88%) and ee's (>99.9%) from the corresponding keto acid salt precursor by a continuous enzymatic reduction in a stirred tank reactor (Tao and McGee 2004). The chemoenzymatic route was signifi-

Candida antarctica lipase B, 40% acetone; racemic cis/trans mixture (CO_2Et) → (CO_2H) >98.5% ee, 49% yield → GABA inhibitor

Scheme 3. Enzymatic resolution of four stereoisomers and dramatic solvent effects

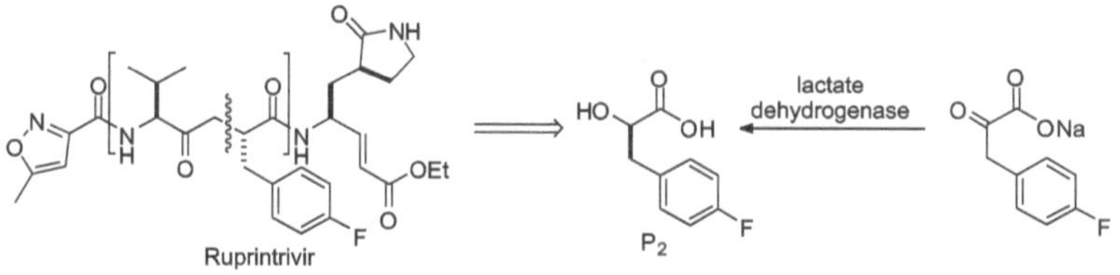

Scheme 4. Chemoenzymatic synthesis of Ruprintrivir™ by enzymatic reduction

cantly shorter than the previous chemical process resulting in significant cost savings.

Another strategy for integrating chemistry and biocatalysis is to modify the substrates chemically by introducing different protecting groups without sacrificing overall synthetic efficiency. In many cases, a particular protecting group may turn out to be the optimal substrate for an enzyme leading to high selectivity and/or productivity. To optimize biocatalytic reactions, substrate modulation might offer a faster and economic alternative to the directed evolution approach.

For example, a standard bond disconnection of a HIV protease inhibitor led to two key precursors: AHPBA (3-amino 2-hydroxybutyric acid) and DMDFP (N-Boc 3,3-dimethyl 4,4-difluoroproline) (Scheme 5).

Enantioselective chemical synthesis of this unusual amino acid DMDFP is problematic mainly because of the steric bulkiness caused by the ring substitution and the difluoro unit. Under enzymatic hydrolysis of the ester substrate (Scheme 6), only *subtilisin Carlsberg* was identified to be active from a library of over 150 hydrolases. However, the reaction was slow, requiring 4–5 days at a low substrate loading of 10 g/l, rendering the process impractical at large scale. By introducing a benzyl protecting group instead of the Boc, pig liver esterase (PLE) was selected from the same enzyme library, which catalyzes the reaction with excellent resolution yield (45%) within 24 h at a substrate loading of 100 g/l. The benzyl group can be removed by hydrogenolysis in downstream steps without introducing additional synthetic steps (Hu et al. 2006). Directed evolution of *subtilisin Carlsberg* can be challeng-

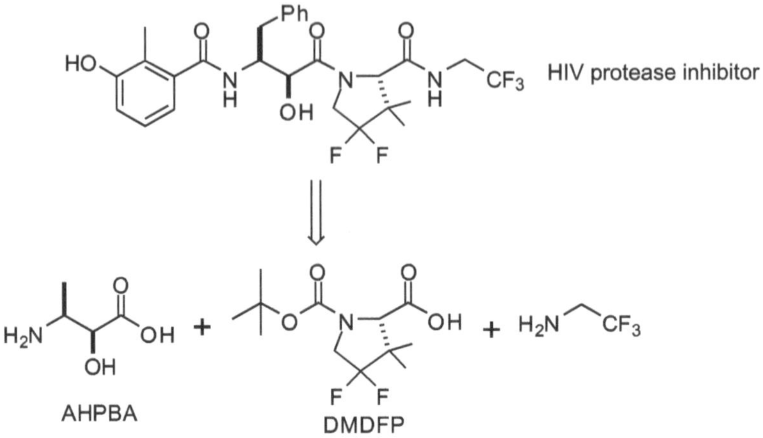

HIV protease inhibitor

AHPBA DMDFP

Scheme 5. Synthesis of a HIV protease inhibitor

P = Boc, 10 g/L using *Subtilisin carlsberg*
P = Bn, 100 g/L using PLE

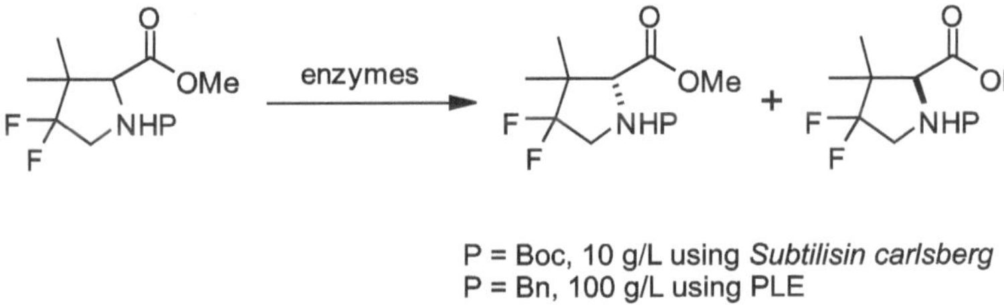

Scheme 6. Substrate modulation in enzymatic resolution of a proline ester

ing and time-consuming because of the intrinsic toxicity of this protease in *E. coli*.

Substrate modulation can also be a good strategy to expand the synthetic applications of biotransformations. For example, enzymatic resolution of secondary amines is often problematic by either hydrolysis of the corresponding amides in water or acylation of free amines in organic solvents. However, using an oxalamic ester as the labile protecting group, a variety of amines can be resolved with excellent enantioselectivity and reactivity through enzymatic hydrolysis (Scheme 7) (Hu et al. 2005).

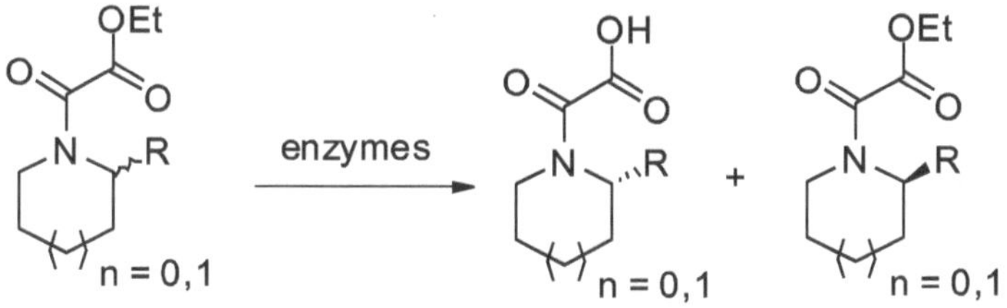

Scheme 7. Chemoenzymatic preparation of optically pure secondary amines via hydrolysis of an racemic oxalamic ester linker

In many cases, retrosynthetic analysis, substrate modulation, and solvent engineering can be integrated synergistically. One example is the synthesis of GARFT inhibitor Pelitrexol™. The first-generation chemical process required 20 steps with an overall yield of only 2%. By integrating with biotransformations, retrosynthetic analyses led to a new route combining Sonogashira coupling with an enzymatic resolution of the resulting adduct (Scheme 8) (Hu et al. 2006).

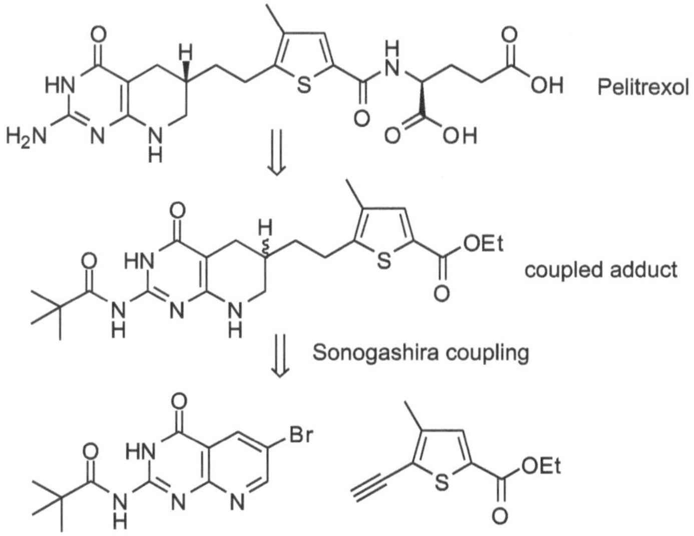

Scheme 8. Retrosynthetic analysis of Pelitrexol™

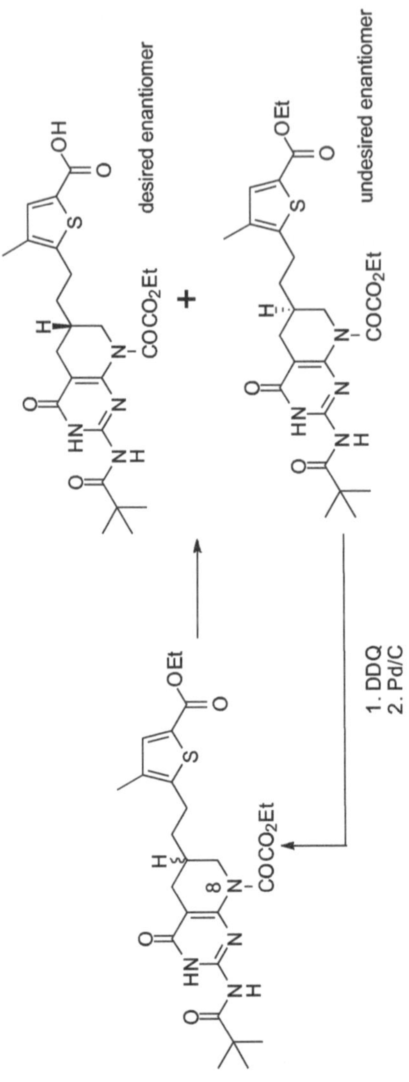

Scheme 9. Enzymatic synthesis of Pelitrexol™ intermediate

It should be noted that asymmetric chemical synthesis of the coupled adduct has proved to be exceedingly difficult despite extensive efforts. In this molecule, the chiral center is six bonds away from the ester group and not surprisingly direct enzymatic hydrolysis of the substrate gave low selectivity. However, when a labile oxalamic ester protecting group was introduced for the nitrogen at the eighth position, high ee's (95%) were obtained with good resolution efficiency (100 g/l substrate, 45% conversion within 4–5 h) (Scheme 9) (Hu et al. 2006). The use of 30% DMF is crucial to this process, which increases both the solubility of the substrate in water and the enantioselectivity of the enzyme. It is worth noting that the enzymatic hydrolysis takes place at the thiophene ester rather than the oxalamic ester. The oxalamic ester group was readily removed in the subsequent workup. Moreover, the wrong enantiomer can be recycled under oxidation and hydrogenation conditions. In this fashion, the new process required only nine steps and the overall yield was increased to 10%–15% yield from the original 2%.

In addition to the optimization of substrates and medium, careful investigation of enzyme kinetics and inhibition is often crucial for the development of a cost-effective and robust process. For example, in an enzymatic synthesis of (*S*)-CNDE (cyanodiester) (Scheme 10), a potential key intermediate for the preparation of pregabalin, the active ingredient of Lyrica®, the initial maximum substrate loading of the racemic CNDE is low (Hu et al. 2005). Kinetic studies show there is significant product inhibition. Screening of a variety of organic and inorganic counter-cations led to the discovery that the production inhibition can be removed by the addition of a catalytic amount of calcium salt in the reaction. Subsequently, the substrate loading was improved several fold, resulting in a practical, cost-effective and green chemoenzymatic process suitable for commercialization (Hu et al. 2005). The wrong enantiomer can be readily recycled (not shown).

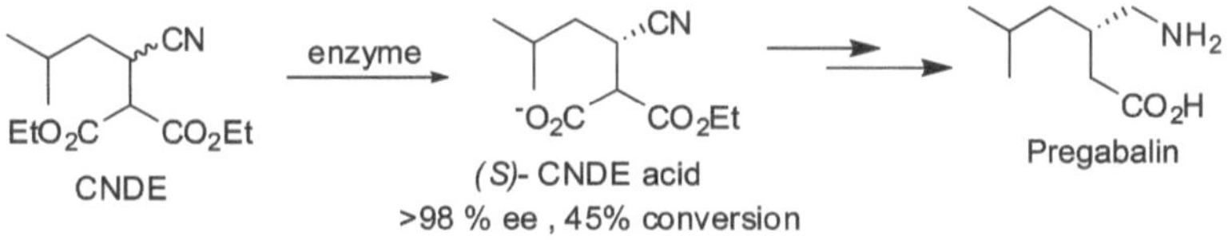

CN
EtO_2C CO_2Et
CNDE
enzyme
CN
^-O_2C CO_2Et
(*S*)- CNDE acid
>98 % ee , 45% conversion
NH_2
CO_2H
Pregabalin

Scheme 10. Chemoenzymatic synthesis of pregabalin by overcoming product inhibition

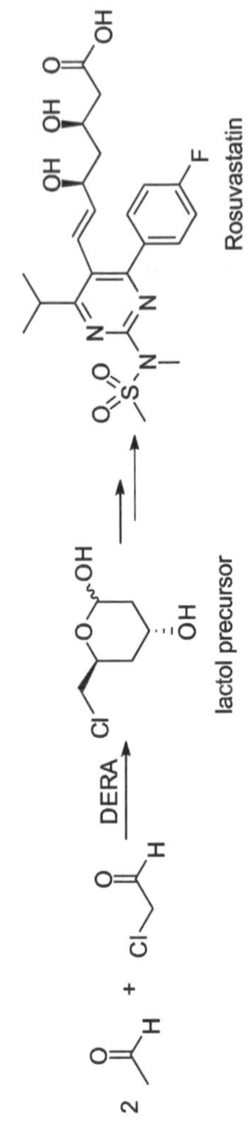

2 CH3CHO + ClCH2CHO —DERA→ lactol precursor → → Rosuvastatin

Scheme 11. Aldolase-catalyzed synthesis of statin drugs via the megagenomic approach

Novel enzyme discovery and directed evolution shall be part of the solution of an integrated approach to developing chemoenzymatic processes. Novel enzymes can also be discovered from environmental samples as illustrated in the synthesis of a lactol precursor for statin-type HMG-CoA reductase inhibitors, including rosuvastatin and atorvastatin (not shown) (Scheme 11). The initial process using an *E. coli* deoxyribose-5-phosphate aldolase (DERA) suffered from high enzyme loading (20% wt/wt) and low volumetric productivity (2 g/l per day). By a combination of activity- and sequence-based screening, a novel DERA was discovered from environmental DNA libraries using the metagenomic approach. With the new aldolase, the volumetric productivity was improved to 30 g/l per hour and the catalyst load was reduced to 2% wt/wt (Greenberg et al. 2004).

Another example of applying directed evolution is the development of a nitrilase-catalyzed synthesis of (*R*)-4-cyano-3-hydroxybutyric acid, a key intermediate in the current manufacturing process of atorvastatin (Scheme 12). For the wild-type nitrilase identified from the initial screening, (*R*)-4-cyano-3-hydroxybutyric acid was obtained with an ee of 88% from the conversion of 3-hydroxyglutaronitrile (3M) (DeSantis et al. 2002). Gene site saturation mutagenesis was then used to improve the lead enzyme and a resulting Ala190His mutant has an enantioselectivity of 99% under the same substrate loading (DeSantis et al. 2003; Bergeron et al. 2006).

To screen a large library of mutants, directed evolution is often much more time-consuming and costly than reaction engineering, substrate modulation, solvent screening, and kinetics studies. It is therefore desirous to build a high-quality, focused library and develop a robust high-throughput screening protocol to reduce the time and cost to identify the best mutant (Reetz et al. 2006; Fiet et al. 2006). One efficient approach is to build a homologous 3D model for an enzyme, whose structure is unknown, from an existing structure solved by crystallography or NMR.

As sequences in the gene bank grow rapidly, a small but diverse library of enzymes can be generated in a short period of time using a bioinformatic approach. For example, in the development of a nitrilase-catalyzed synthesis of pregabalin, the API of Lyrica® (Burns et al. 2005) (Scheme 13), a targeted nitrilase library was discovered from the gene bank, allowing the identification of an initial hit, a nitrilase from *Ara-*

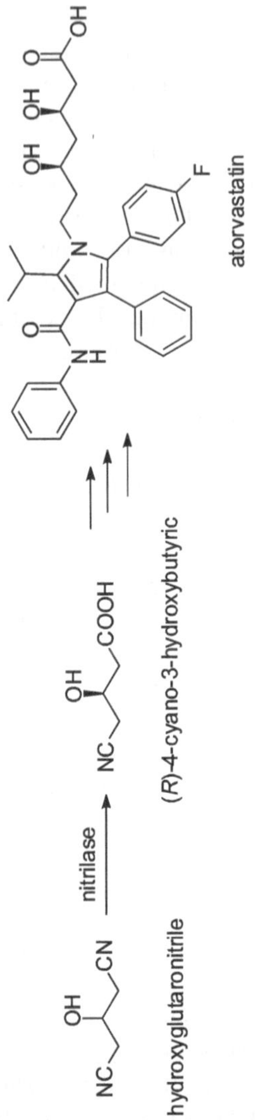

hydroxyglutaronitrile

nitrilase

(*R*)-4-cyano-3-hydroxybutyric

atorvastatin

Scheme 12. Nitrilase-catalyzed synthesis of atorvastatin intermediate by directed evolution

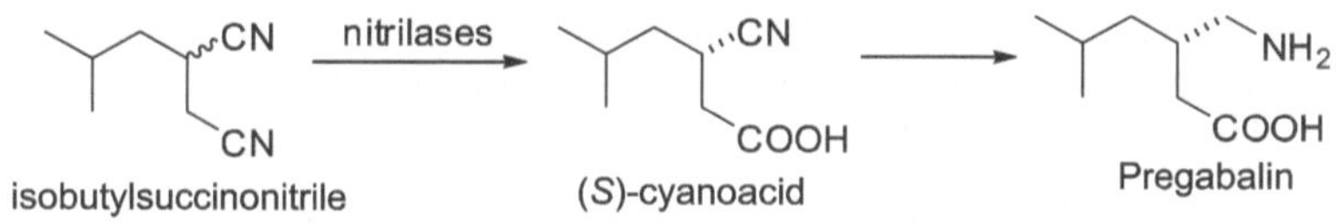

Scheme 13. Nitrilase-catalyzed synthesis of pregabalin™ by ePCR

bidopsis thaliana. This enzyme was subsequently engineered by ePCR to improve its reactivity (Yazbeck et al. 2006; Xie et al. 2006).

2 Conclusions

Biotransformation is emerging to be a powerful technology for chemical synthesis and uniquely suited for the development of cost-effective and environmentally friendly solutions for drug manufacture (Tucker 2006). The successful implementation of biotransformations at industrial scale requires the strategic use of medium screening, substrate modulation, reaction engineering, enzymology, and discovery and evolution of biocatalysts. Most importantly, biocatalysis and modern chemical research and development need to be integrated at the retrosynthetic level to deliver efficient and practical routes with fewer synthetic steps and significantly reduced waste streams (Tao et al. 2006).

References

Bergeron S, Chaplin D, Edwards JH, Ellis BS, Hill CL, Holt-Tiffin K, Knight JR, Mahoney T, Osborne AP, Ruecroft G (2006) Nitrilase-catalyzed desymmetrization of 3-hydroxyglutaronitrile: preparation of a statin side-chain intermediate. Org Proc Res Dev 10:661–665

Burns M, Weaver J, Wong J (2005) Stereoselective enzymic bioconversion of aliphatic dinitriles into cyano carboxylic acids. WO 2005100580

DeSantis G, Zhu Z, Greenberg W, Wong K, Chaplin J, Hanson SR, Farwell B, Nicholson LW, Rand CL, Weiner DP, Robertson D, Burk MJ (2002) An enzyme library approach to biocatalysis: development of nitrilases for enantioselective production of carboxylic acid derivatives. J Am Chem Soc 124:9024–9025

DeSantis G, Wong K, Farwell B, Chatman K, Zhu Z, Tomlinson G, Huang H, Tan X, Bibbs L, Chen P, Kretz K, Burk MJ (2003) Creation of a productive, highly enantioselective nitrilase through gene site saturation mutagenesis (GSSM). J Am Chem Soc 125:11476–11477

Dragovich PS, Webber SE, Babine RE et al (1998) Structure-based design, synthesis, and biological evaluation of irreversible human rhinovirus 3C protease inhibitors. J Med Chem 41:2819–2834

Fiet SVS, Beilen JBV, Witholt B (2006) Selection of biocatalysts for chemical synthesis. Proc Natl Acad Sci USA 103:1693–1698

Greenberg WA, Varvak A, Hanson SR, Wong K, Huang H, Chen P, Burk MJ (2004) Development of an efficient, scalable, aldolase-catalyzed process for enantioselective synthesis of statin intermediates. Proc Natl Acad Sci USA 101:5788–5793

Hu S, Tat D, Martinez C, Yazbeck D, Tao J (2005) An efficient and practical chemoenzymatic method for preparing optically active secondary amines. Org Lett 7:4329–4331

Hu S, Martinez CA, Tao J, Tully W, Kelleher P, Dumond Y (2005) An efficient method for the production of enantiopure pregabalin. US Pat Appl Publ US 2005283023

Hu S, Martinez CA, Tao J, Tully W, Kelleher P, Dumond Y, McLoughlin M (2005) Green chemistry in the redesign of pregabalin processes, 2nd International Conference on Green and Sustainable Chemistry and 9th Annual Green Chemistry & Engineering Conference, Washington, DC, June 20–24

Hu S, Martinez C, Kline B, Yazbeck D, Tao J, Kucera D (2006) Efficient and scalable enzymatic process for the production of *(2S)*-4,4-difluoro-3,3-dimethyl-N-Boc-proline, a key intermediate in the synthesis of HIV protease inhibitors. Org Process Res Dev 10:650–654

Hu S, Kelly S, Lee S, Tao J, Flahive E (2006) Efficient chemoenzymatic synthesis of Pelitrexol via enzymatic differentiation of a remote stereocenter. Org Lett 8:1653–1655

Leresche JE, Meyer HP (2006) Chemocatalysis and biocatalysis (biotransformation): some thoughts of a chemist and of a biotechnologist. Org Process Res Dev 10:572–580

Martinez C, Yazbeck D, Tao J (2004) An efficient enzymatic preparation of rhinovirus protease inhibitor intermediates. Tetrahedron 60:759–764

Reetz MT, Wang LW, Bocola M (2006) Directed evolution of enantioselective enzymes: iterative cycles of CASTing for probing protein-sequence space. Angew Chem Int Ed 45:1236–1241

Straathof AJ, Panke S, Schmid A (2002) The production of fine chemicals by biotransformations. Curr Opin Biotechnol 13:548–556

Tao J, McGee K (2002) Development of a continuous enzymatic process for the preparation of (R)-3-(4-fluorophenyl)-2-hydroxypropionic acid. Org Process Res Dev 4:520–524

Tao J, McGee K (2004) Chapter in asymmetric catalysis on industrial scale. Wiley-VCH, Weinhein, Germany, pp 323–334

Tao J, Pettman A, Liese A (2006) Chapter in industrial biotransformations. Wiley-VCH, Weinhein, Germany, pp 63–91

Tucker JL (2006) Green chemistry, a pharmaceutical perspective. Org Proc Res Dev 10:315–319

Xie J, Feng J, Garcia E, Bernett M, Yazbeck D, Tao J (2006) Cloning and optimization of a nitrilase for the synthesis of (3*S*)-3-cyano-5-methyl hexanoic acid. J Mol Cat B, in press

Yazbeck D, Tao J, Martinez C, Kline B, Hu S (2003) Automated enzyme screening methods for the preparation of enantiopure pharmaceutical intermediates. Adv Syn Cat 4:524–532

Yazbeck D, Martinez C, Hu S, Tao J (2004) Challenges in the development of an efficient enzymatic process for the pharmaceutical industry. Tetrahedron Asym 15:2757–2763

Yazbeck D, Derrick A, Panesar M, Deese A, Gujral A, Tao J (2006) Enzymatic process for the synthesis of *Z/E*-(2*R*, 5*R*)-bicyclo[3.2.0]hept-6-ylidene-acetate: solvent effect and NMR study. Org Proc Res Dev 10:655–660

Yazbeck D, Durao P, Xie J, Tao J (2006) A high-throughput metal ion-based method for nitrilase screening. J Mol Cat B 39:156–159

Ernst Schering Foundation Symposium Proceedings, Vol. 3, pp. 133–149
DOI 10.1007/2789_2007_032

Published Online: 23 May 2007

Scale-Up in Microwave-Accelerated Organic Synthesis

H. Lehmann(✉)

Discovery Technologies/Preparation Laboratories,
Novartis Institute for Biomedical Research, Klybeckstrasse 191, 4002 Basel, Switzerland
email: *hansjoerg.lehmann@novartis.com*

Abstract. Microwave-assisted organic chemistry has received strong exposure in the literature over the last decade, and nowadays more and more research chemists are successfully applying microwave technology to organic reactions on a small scale. However, the efficient application of this technology to cover the specific needs of larger-scale preparations, e.g., in a kilo lab, remains to be shown. We therefore initiated a study to investigate the scalability of microwave technology. Two different microwave systems designed for large-scale operation were evaluated in order to characterize strengths and weaknesses of each instrument with regard to scale-up. Special focus was directed on temperature/pressure limits, handling of suspensions, ability to rapidly heat and cool, robustness, and overall processing time. Based on the results of this study, a batch microwave reactor with a reaction volume of approximately 1.1 l was purchased and installed in the kilo lab. Several reactions have been performed successfully on a 50- to 100-g scale in our laboratory, showing that a scale-up from a 15 ml scale to a 1-l scale is feasible. In general, a significant reduction of reaction time was achievable, in some cases yields and selectivity were also improved. Nevertheless, a major weakness of the available systems is the limited vessel size, which is, in most cases, far below a suitable reaction volume required for work in a kilo lab.

1 Introduction

During the last decade, requirements on modern medicinal chemistry have led to the development of new strategies and technologies. In order to identify more potential drug candidates, pharmaceutical companies have made major investments in high-throughput technologies, combinatorial chemistry and synthesis of huge new compound libraries. In consequence, increasing interest has been directed toward new technologies that allow more rapid synthesis of new chemical compounds. Microwave-assisted heating under controlled conditions has proved to be an invaluable technology for medicinal chemistry, since it often dramatically reduces reaction times and gives access to reaction conditions that are not attainable under conventional thermal heating (Kappe 2005; Nicewonger et al. 2003). Microwave heating offers the opportunity to save time in the lab and to improve productivity. Research chemists at Novartis are therefore increasingly using microwave reactors to carry out chemical reactions on a small scale. In recent years, approximately a dozen microwave devices have been installed at Novartis in medicinal chemistry and even more will be acquired in the near future. A fully automated high-throughput microwave synthesis factory consisting of four microwave reactors has been installed that is currently able to perform thousands of reactions within a week. It is obvious that in the future this new technique will become a standard technology in research laboratories because of its success in small-scale organic synthesis, in particular for the rapid optimization of reaction conditions (Kappe 2004). As a consequence of these developments, there is an increasing pressure, in the context of the drug discovery process, to use this technology also on larger scale, for example in a kilo lab. Nevertheless, published examples of scale-up experiments are rare to date, in particular those involving complex organic reactions (Kappe et al. 2003). This is coupled with the fact that only a few companies are currently developing microwave devices for scale-up.

One goal of our investigation was to evaluate the potential of commercially available microwave technology on a large scale (at least 300–500 ml reaction volume) in order to reduce the processing time for difficult reactions requiring long reaction times under conventional conditions. In addition, a second target was to find out which commercially available microwave system best fits to the requirements of scale-up (e.g., maximum reaction volume, easy to use) and what benefit can be expected from using a microwave device (e.g., increased productivity, access to new reaction types) on a large scale.

2 Evaluation Approach

Until now, synthesis work in a kilo lab has been normally performed batchwise using conventional thermal heating techniques under atmospheric pressure, so that the achievable reaction temperature is limited by the boiling point of the solvent. In contrast, microwave heating in a pressurized vessel allows access to reaction temperatures far above the boiling point of the solvent and leads to remarkably shortened reaction times. Also, by choosing the solvent independent of its boiling point but on more favorable work-up and purification properties, time-consuming work-up procedures can be avoided, which should result in a significantly shorter overall processing time for a chemical synthesis.

In considering a collection of model reactions, we intended to compare conventional heating in an oil bath with microwave heating. We first chose a microwave reactor for small scale work: the Biotage Emrys Optimizer (20-ml glass vials, filling volume approximately 15 ml). In a second stage, the same reactions were carried out on a larger scale using a multimode batch reactor (Anton Paar Synthos 3000, 16 × 100-ml PTFE vessels in a ceramic vessel jacket, filling volume 60–70 ml for each vessel, see Fig. 1) and a monomode stop flow reactor (CEM Voyager SF, 80-ml glass vessel, filling volume 50 ml, see Fig. 2). For comparison, some of these reactions were done additionally in an autoclave.

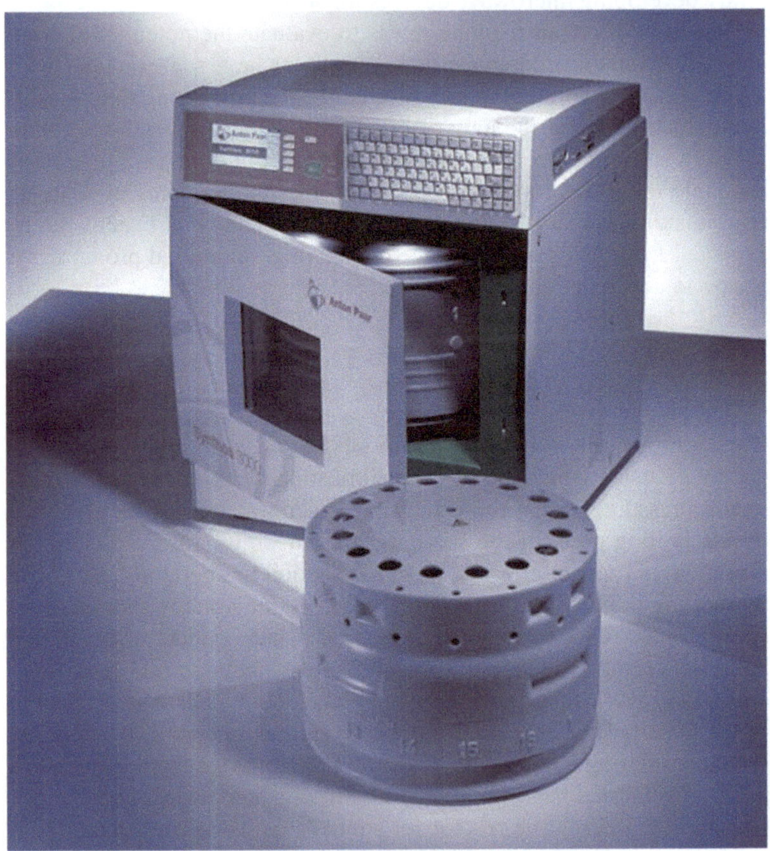

Fig. 1. Synthos 3000 (Anton Paar)

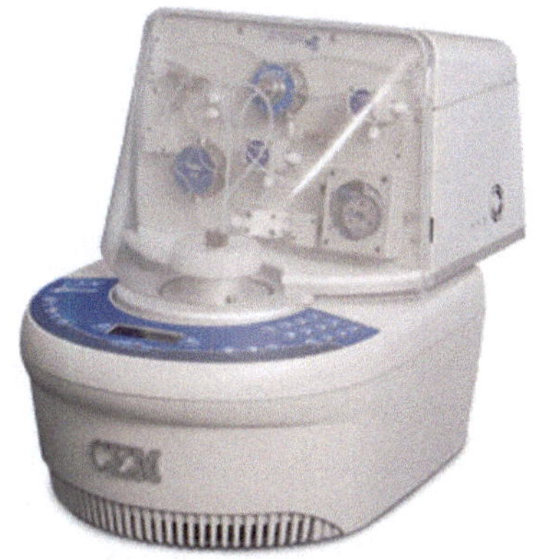

Fig. 2. Voyager SF (CEM)

A range of technical details of the microwave reactors used in this study are shown in Table 1.

Requirements for use of those devices in our kilo lab were defined as follows:

- Synthesis of compounds on a scale greater than 100 g in a reasonable timeframe; use of microwave technology should result in time saving compared to conventional heating.
- No additional optimization work for transfer from the small scale to the large scale.
- Ability to process heterogeneous mixtures, especially suspensions, without problems.
- Reproducibility.
- Ease of handling.

In addition, the possibility to run reactions automatically in a continuous or repetitive manner would also be regarded as an advantage.

On the other hand, it must be kept in mind that a number of critical issues are linked with the application of microwave in chemical scale-up (Ondruschka et al. 2004). Firstly, penetration depth of microwaves in

Table 1 Commercially available microwave reactors used in our evaluation

	Optimizer (Biotage)	Synthos 3000 (Fig. 1)	Voyager SF (Fig. 2)
Vessel size (ml)	5–20	16×100	80
Reaction volume (ml)	15	1,000–1,200	50
Max. pressure (bar)	20	40	17
Max. temp. (°C)	250	240	220
Cooling system	Compressed air	Air ventilation	Compressed air

suitable solvents is limited to 2–5 cm and therefore the vessel size cannot be expanded unrestrictedly. Secondly, high microwave power may lead to rapid input of energy into the reaction and to overheating. Further issues are field homogeneity (inhomogeneity may lead to hot spots and product degradation) and stability of solvents, reagents, and products at temperatures higher than 200°C. Instability and degradation of the reaction mixture may lead to safety problems. The construction of the microwave reactor must be able to withstand high temperatures and high pressure (e.g., 250°C, 20 bar). Furthermore, it is important to measure the temperature inside the reaction mixture because IR sensors outside the reaction vessel always register a significantly lower temperature because of the inverted heat flow from inside to outside.

3 Investigation on Model Reactions

For model reactions, we chose the aromatic substitution of aryl halides with nucleophiles such as phenolates or amines. The reaction parameters particularly focused upon were reaction time, selectivity, work-up procedure, and overall processing time.

3.1 Model Reaction 1

Biaryl ethers are a well-known motif in medicinal chemistry. The reaction of p-nitrophenol with 2,4-dichloropyrimidine was performed in a mixture of acetone and water from which the product precipitated after completion. Work-up and isolation of the product was therefore very simple. We also observed the formation of a by-product (depending of the exact reaction conditions used), a side-reaction that is known in literature. Results of reaction 1, which was performed under various conditions, are summarized in Table 2.

As illustrated in Scheme 1, on a small scale under reflux in acetone/water, 5% of starting material remained after 12 h reaction time and approximately 20% of the by-product was formed (entry 1). When performing the reaction at the same concentration in a lab-scale microwave device (Emrys Optimizer, entry 2) at 120°C, the reaction was complete after 5 min and gave a product of significantly higher purity and in higher yield. In the next step, 400 ml of reaction mixture was reacted in an 8 vessel rotor batch microwave (entry 3) at the same temperature

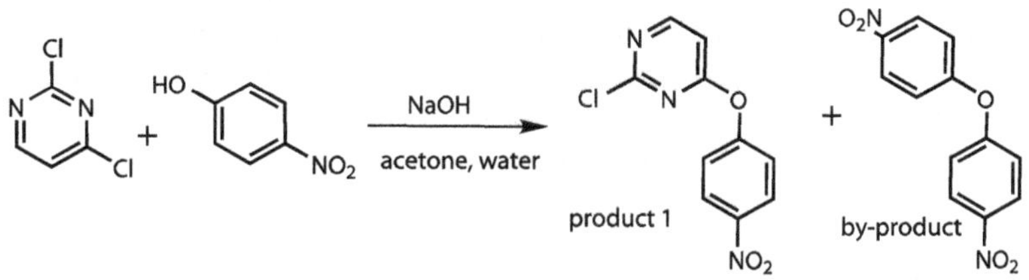

NaOH

acetone, water

product 1

by-product

Scheme 1. Reaction 1

Table 2 Result of reaction 1 under various conditions

Entry		Batch size	Reaction temp., time	Process time	Isol. yield	Purity	Remarks
1	Conv. heating	20 ml	62°C, 12 h	12 h 30 min	66%	80%	5% educt left, 20% by-product
2	μW, Emrys Optimizer	15 ml	120°C, 5 min	10 min	83%	>96%	100% conversion, < 5% by-product
3	μW, Synthos 3000	8×50 ml	120°C, 10 min	30 min	76%	90%	10% by-product
4	μW, Voyager SF	400 ml	120°C, 15 min	240 min	81%	95%	5% by-product, clogged lines
5	Autoclave	400 ml	120°C, 10 min	275 min	74%	85%	>10% by-product

for 10 min. After a cooling down period of 18–20 min, the precipitated product was isolated by simple filtration in 76% yield but also 10% of the by-product was formed under these conditions. The overall processing time, which includes heating, maintaining the temperature for 10 min, and cooling, was only 30 min.

In contrast to this, conversion of a reaction volume of 400 ml in the Voyager SF took roughly 4 h (entry 4) because eight cycles on a 50-ml scale were necessary to convert this volume. Each cycle consisted of adding the starting material, heating, maintaining the temperature for 15 min, cooling, and removing the mixture from the reaction vessel, requiring a total of 30 min. Furthermore, when the mixture was pumped out of the vessel after the reaction, precipitation of the product clogged the outlet tube and a valve, so that the process had to be interrupted manually. For comparison, the reaction was also carried out in an autoclave (1 l volume, entry 5). Because of the long heating and cooling process, processing time in total was 275 min, leading to a product in 74% yield that contained more than 10% of the by-product.

3.2 Model Reaction 2

Scheme 2 shows a substitution reaction which is often carried out with solid carbonate in suspension, which is easily removed after reaction by a simple filtration. This reaction was chosen not only to test the handling of suspensions, but also to investigate the influence of different solvents on selectivity and work-up. The results are shown in Table 3.

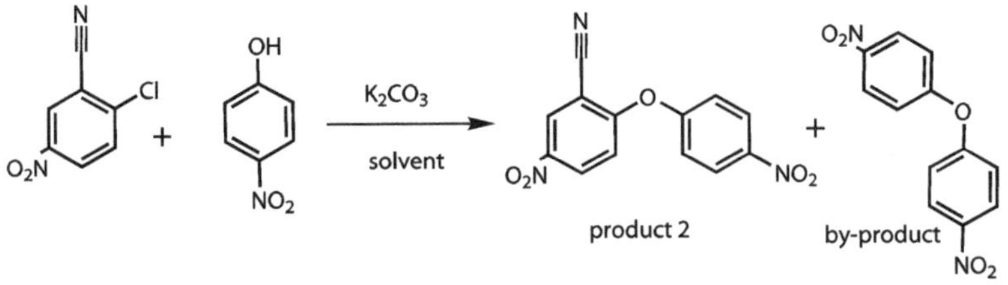

N
Cl
O_2N
+
OH
NO_2
K_2CO_3
solvent
N
O
O_2N
NO_2
product 2
+
O_2N
O
by-product
NO_2

Scheme 2. Reaction 2

Table 3 Result of reaction 2 under various conditions

Entry		Batch size	Solvent temp., time	Process time	Isol. yield	Purity	Remarks
1	Conv. heating	20 ml	CH_3CN, reflux, 6 h	6 h 30 min	76%		dark purple crystals
2	Conv. heating	20 ml	DMSO, 150°C, 10 min	60 min	57%	65%	Mix of product and by-product isolated
3	µW, Emrys optimizer	15 ml	CH_3CN, 150°C, 10 min	17 min	98%	>96%	Yellow crystals
4	µW, Synthos 3000	4×50 ml	CH_3CN, 150°C, 10 min	25 min	78%	>96%	In a second trial 73% yield isolated
5	µW, Voyager SF	300 ml	CH_3CN, 150°C, 10 min				Inlet tube clogged
6	Autoclave	200 ml	CH_3CN, 150°C, 10 min	280 min	92%	92%	8% by-product

Under conventional conditions in the lab (acetonitrile, reflux for 60 min) the product was isolated in 86% yield but starting material remained, revealing that the reaction was not complete. Prolongation of the reaction time to 6 h led to full conversion of the starting material, but also to reduced yield and to dark purple crystals (entry 1). Performing the same reaction in DMSO at higher temperature (150°C for 10 min) led to a significantly shorter reaction time, but also to increased formation of by-product and to a significant lower yield (entry 2). In contrast, when a microwave reactor was used at 150°C for 10 min, an isolated yield of 98% was achieved (entry 3). Scale-up trials then were done in the Synthos 3000, where the suspension was reacted without problems. In order to check reproducibility, two identical batches were carried out: the first batch giving 78% yield (entry 4) and the second giving 73% yield after crystallization. When the same reaction was done in an autoclave, a grey product, containing 8% of by-product, was isolated in

92% yield but with low quality (entry 6). Using the Voyager SF for this reaction failed, because the system was not able to pump the suspension into the reaction vessel (entry 5). Although we used finely ground potassium carbonate, the inlet tube was repeatedly blocked by the solid carbonate. It should also be mentioned that in terms of processing time, the reaction in an autoclave required 280 min, whereas in the Synthos 3000 only 25 min were needed to do the conversion.

3.3 Model Reaction 3

Scheme 3 illustrates the third model reaction was homogenous, and again we intended to compare conventional heating with microwave heating, as well as study the influence of different solvents on both. The obtained results are shown in Table 4.

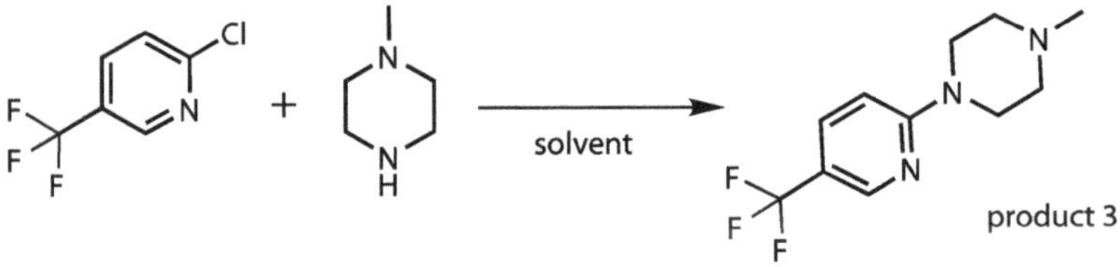

solvent

product 3

Scheme 3. Reaction 3

Two different typical work-up procedures were applied in order to demonstrate the influence on the yield.

Work-up A: The reaction mixture was poured into water and extracted with DCM. The organic phase was evaporated and crude product was purified by flash chromatography.

Work-up B: The reaction mixture was poured into water. The precipitated product was filtered, washed, and dried.

Under reflux conditions on a small scale in acetonitrile, the reaction was complete only after 24 h (entry 1). In a microwave reactor at 150°C, 86% isolated yield was achieved after 20 min reaction time (entry 2). Running the same reaction in DMSO or DMF at 150°C gave complete conversion of starting material, but because of losses during work-up, isolated yield was lower (entries 3 and 4). For scale-up trials, the Voyager SF was used, and the reaction was performed at 180°C for 20 min

Table 4 Results of reaction 3

Entry		Batch size	Solvent temp., time	Process time	Isol. yield	Purity	Remarks
1	Conv. heating	20 ml	CH_3CN, reflux, 24 h		97%	>97%	Reaction completed, work-up A
2	μW, Emrys optimizer	15 ml	CH_3CN, 150°C, 20 min	28 min	86%	>97%	Reaction not completed, work-up A colorless crystals
3	Conv. heating	20 ml	DMF, 150°C, 20 min	50 min	60%	>99%	Work-up B brownish crystals
4	Conv. heating	20 ml	DMSO, 150°C, 20 min	50 min	63%	>99%	Work-up B beige crystals
5	μW, Voyager SF	5×50 ml	CH_3CN, 180°C, 20 min	2 h 40 min	72%	95%	Work-up B colorless crystals

in each cycle. To process a 250 ml batch, five cycles of 50 ml each had to be performed, leading to a processing time of 2 h 40 min (entry 5). A typical diagram of temperature, pressure and time of all five runs is shown in Fig. 3. The temperature rises very rapidly to 180°C and is held for 20 min, the pressure during the reaction is approximately 11 bar. Then the vessel is cooled by air ventilation to about 60°C within 8–10 min. In total, this results in a cycle time of about 32 min for each cycle.

3.4 Discussion of the Results from Model Reactions

Two different microwave reactors for scale-up have been tested with special focus on handling, automation, workflow, and typical scale-up issues. The Synthos 3000 as a batch mode reactor provides a relatively large reaction volume (16 × 70 ml), allows high temperature and pressure (240°C/40 bar), and proved to be very robust. Handling of suspen-

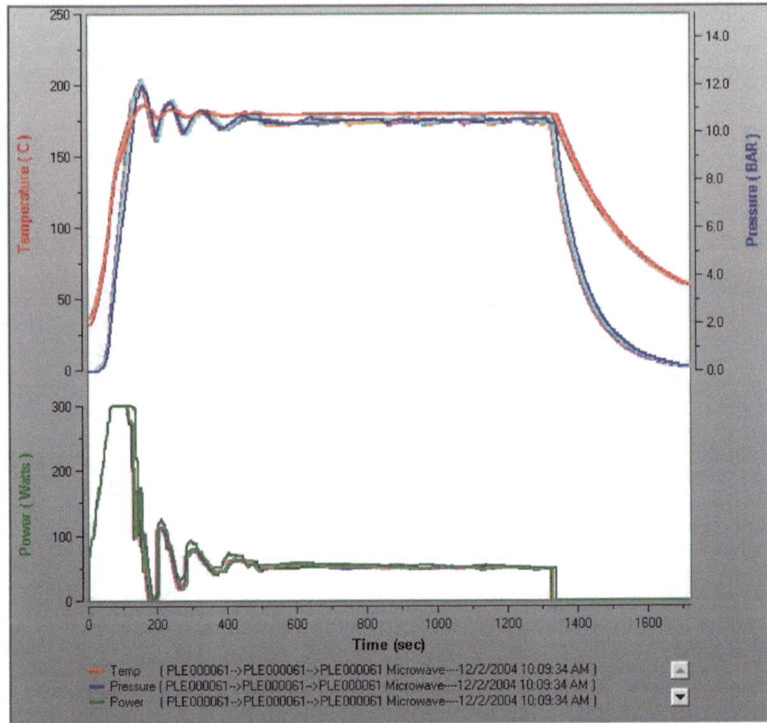

Fig. 3. Diagram of reaction 3, Voyager SF

sions gave no problems and the system was shown to have very good temperature control. The software is easy to handle and results from the small scale were reproduced on a larger scale without additional optimization effort. Nevertheless, for synthesis on a large scale, filling and removing the contents of all 16 vessels is laborious and the cooling period is comparatively long. The reactor tested does not allow the handler to run several reactions sequentially in an automated manner. Automation of this type would be a highly desirable option in order to optimize the work-flow in a kilo lab.

The possibility for automation is offered by the Voyager SF with its stop flow system, able to process larger reaction volumes by sequentially repeating reactions on a 50-ml scale. The relatively small reaction volume guarantees a low safety risk on one hand, but on the other hand leads to relatively long processing times. Handling of suspensions also proved to be a critical issue and, in one case, precipitated product blocked the outlet tube and valves. Running reactions for a longer period without monitoring, for example overnight, can only be therefore recommended for homogeneous reaction mixtures.

4 Examples for Scale-Up Using Microwaves

In the course of our daily business in the kilo lab, we are faced with the preparation of a number of intermediates or drug candidates on a larger scale in which at least one step of the original synthesis was performed in a microwave reactor. In order to increase the productivity, it is of great interest to reduce the time needed to produce these compounds on a large scale. Reactions taking a long time at elevated temperature are targets for improvement in work-flow using microwave technology.

4.1 Example 1

In Scheme 4 it can be seen that cyclisation of the ketoester with hydrazine-hydrate was part of a reaction sequence given to the kilo lab and is described in literature with a yield of approximately 60% (Schenker and Salzmann 1979). Under conventional conditions (refluxing in ethanol for 4 h), we obtained the product in 53% yield. Performing the same reaction in the Synthos 3000 microwave reactor at 140°C for 20 min, we improved the yield to 64% on a 20-g scale. Further scale-up to a batch size of 130 g of product was performed under the same conditions (140°C/20 min) with no significant problems.

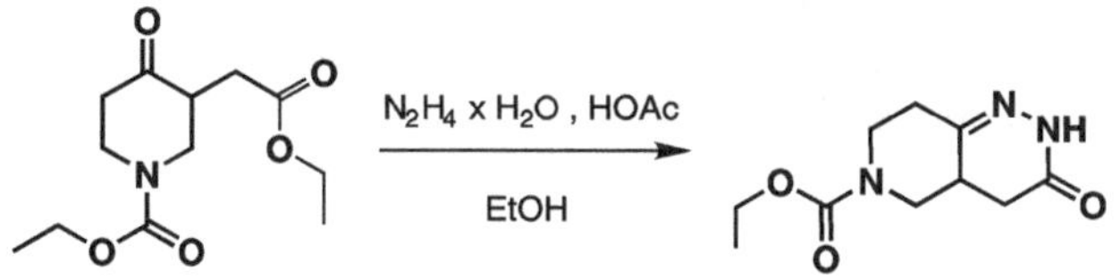

$N_2H_4 \times H_2O$, HOAc

EtOH

Scheme 4. Reaction 4

4.2 Example 2

A well known standard procedure in organic chemistry is to form pyrazoles via the reaction of β-ketonitriles with hydrazines. The reaction shown in Scheme 5 is part of the synthesis of a p38 MAP Kinase Inhibitor from Boehringer Ingelheim. In the literature, this reaction is described as giving a yield of 60%–85% after reflux in toluene for 24 h (Regan et al. 2002). Instead of toluene, we used methanol as a solvent and obtained a yield of 92% after 3 h of reflux. Using microwave conditions (120°C, 15 min), we scaled up the reaction from the 20-g scale to the 65-g scale, obtaining 86% yield and reducing the reaction time significantly from hours to minutes.

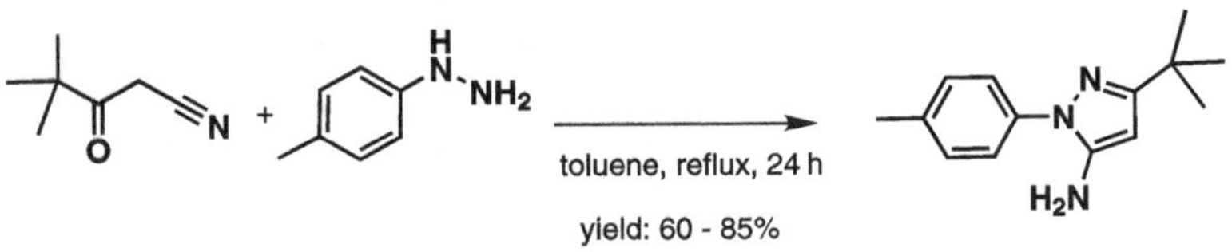

toluene, reflux, 24 h

yield: 60 - 85%

Scheme 5. Reaction 5

4.3 Example 3

Substitution of halogens on heteroaromatic rings is a common way to introduce new functionalities. The product from reaction 6 (Scheme 6) was required on a 100-g scale as an intermediate. In the literature, this exchange was done on a 5-g scale using ammonia in ethanol in a sealed tube under pressure for 6 h at 125–130°C with a yield of 76% (Bendich et al. 1954). Because of the lack of a suitable autoclave for high-pressure reactions, we choose the microwave reactor for scale-up trials. Using our Synthos 3000 equipment, we found suitable conditions with only minimal optimization at 170°C for 180 min and obtained the desired product on a 60-g scale in 83% yield.

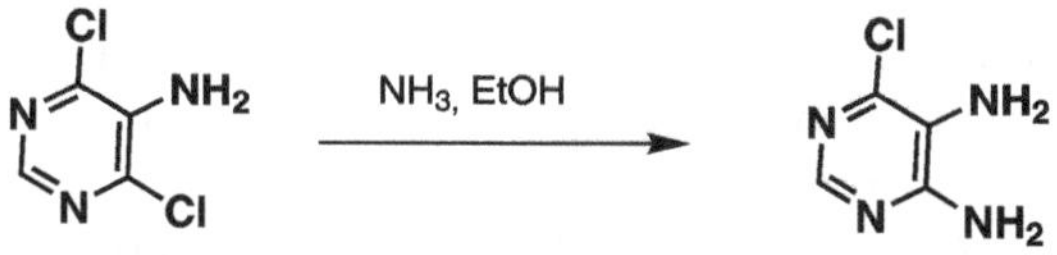

NH_3, EtOH

Scheme 6. Reaction 6

5 Summary

The current study was initiated to evaluate the scalability of microwave technology. The prerequisites for a productive application of microwave-assisted synthesis on a larger scale in a kilo lab were defined, and a market survey for suitable microwave reactors from commercial suppliers was performed. In a second step, model reactions were selected and microwave systems designed for large-scale operation (Synthos 3000 and Voyager SF) were evaluated under realistic conditions in the kilo lab. Special focus was on temperature/pressure limits, handling of suspensions, robustness toward different kinds of chemistry, ability for fast heating and cooling, and the overall processing time. For comparison, some reactions were also performed in an autoclave. Strengths and weaknesses of each instrument were evaluated.

In various experiments, it was shown that the use of microwave technology leads to a significant decrease in the reaction time and in some cases also to less by-product and a higher yield. This technology allowed us to optimize the reaction with focus on work-up and purification, independent of reaction temperature and boiling point of the solvent. In most cases the reaction conditions, applied on 15 ml scale in the Emrys Optimizer, were transferred without further optimization to the microwave reactors tested and led to comparable results. Additional optimization in a few cases was limited to small adjustments in reaction temperature or reaction time. A number of scale-up experiments were conducted using a Synthos 3000 reactor and we showed that a scale-up to 100 g is feasible.

A major weakness of the available systems, however, is the limited vessel size, which in most cases is far below the volume required in a kilo lab of at least 1 l.

In theory, conditions in a microwave reactor, with regard to temperature and pressure, should also be attainable by using an autoclave. Some reactions were therefore done in an autoclave for comparison. The main difference proved to be the longer processing time, mainly caused by the long heating and cooling period when using an autoclave. This leads, as shown in example 2 of our study, to a different selectivity and to a significantly higher amount of by-product, which as a consequence had to be separated by additional purification steps. In this case, using microwave heating is obviously more advantageous than using an autoclave.

In general, it should be noted that the commercially available microwave devices for scale-up were developed either from older existing systems for digestion (e.g., Synthos 3000, Anton Paar) or were completely developed new for synthesis on a larger scale (e.g., Advancer, Biotage). Some of these tools are more or less in a prototype phase where further development is ongoing. As this study revealed, the first steps of microwave technology toward scale-up in larger than the-100 g scale are complete, but in order for this to become a standard technology for kilogram-scale synthesis, further improvements with regard to reaction volume and automation will need to follow in the future

Acknowledgements. The author thanks Biotage (Uppsala, Sweden), Anton Paar (Graz, Austria), and CEM Corp. (Matthews, NC, USA) for their support and for the possibility to test their microwave reactors in the kilolab. He also thanks all the colleagues and people who were involved in this project for their input and the challenging discussions.

References

Bendich A, Russell P, Fox J (1954) The synthesis and properties of 6-chloropurine and purine. JACS 76:6073–6077

Kappe O, Stadler A (2005) Microwaves in organic and medicinal chemistry. Whiley-VCH, Weinhein, Germany

Kappe O (2004) Controlled microwave heating in modern organic synthesis. Angew Chem Int Ed 43:6250–6284

Kappe O, Stadler A, Dallinger D (2003) Scalability of microwave-assisted organic synthesis. Organic Process Res Dev 7:707–716

Nicewonger R, Baldino C, Evans M (2003) The accelerated development of an optimized synthesis of 1,2,4-oxadiazoles. Tetrahedron Lett 44:9337–9341

Ondruschka B, Nüchter M, Bonrath W (2004) Microwave assisted synthesis—a critical technology overview. Green Chem 6:128–141

Regan J, Breitfelder S Cirillo P (2002) Pyrazole urea-based inhibitors of p38 MAP kinase: from lead compound to clinical candidate. J Med Chem 45:2994–3008

Schenker E, Salzmann R (1979) Antihypertensive action of bicyclic 3-hydrazinopyridazines. Arzneimittel-Forschung 29:1835–1843

Reading List

Anton Paar Corporate Website (2007) www.anton-paar.com. Cited 5 February 2007

Biotage (2007) www.biotage.com. Cited 5 February 2007

Cemsynthesis (2006) www.cemsynthesis.com. Cited 5 February 2007

Kappe O (2007) Christian Doppler Institute for Microwave Chemistry, University of Graz, www.maos.net. Cited 5 February 2007

Milestone Microwave Laboratory Systems (2003) www.milestonesci.com. Cited 5 February 2007

Ernst Schering Foundation Symposium Proceedings, Vol. 3, pp. 151–185
DOI 10.1007/2789_2007_033

Published Online: 24 May 2007

Solid Supported Reagents in Multi-Step Flow Synthesis

I.R. Baxendale, S.V. Ley(✉)

Innovative Technology Center (ACS), Department of Chemistry,
University of Cambridge, CB21EW Cambridge, UK
email: *svl1000@cam.ac.uk*

Abstract. The frequently overlooked benefits that considerably simplify and enrich our standard of living are most often hinged upon chemical synthesis. From the development of drugs in the ongoing fight against disease to the more aesthetic aspects of society with the preparation of perfumes and cosmetics, synthetic chemistry is the pivotally involved science. Furthermore, the quality and quantity of our food supply relies heavily upon synthesised products, as do almost all aspects of our modern society ranging from paints, pigments and dyestuffs to plastics, polymers and other man-made materials. However, the demands being made on chemists are changing at an unprecedented pace and synthesis, or molecular assembly, must continue to evolve in response to the new challenges and opportunities that arise. Responding to this need for improved productivity and efficiency chemists have started to explore new approaches to compound synthesis. Flow-based synthesis incorporating solid supported reagents and scavengers has emerged as a powerful way of manipulating chemical entities and is envisaged to become a core laboratory technology of the future.

1 Introduction

Flow-based applications of chemical synthesis have existed for many decades (Karnatz and Whitmore 1932; Dye et al. 1973). Unfortunately, although the concepts of increased mixing efficiency, controlled scaling factors, enhanced safety ratings and continuous processing capabilities have all been well recognised, these benefits have not been generically leveraged into conventional synthetic chemistry. Indeed, traditionally almost all chemical manipulations have been conducted as sequences of well-defined and laboriously optimised single-step transformations operated in batch mode. Consequently, the majority of process innovations in terms of new tools and techniques that have helped define modern organic synthesis have been based on modifications to existing batch-processing protocols. It is certainly true that some of these technological changes have had a significant impact on the way chemists approach and prepare their compound collections. One such methodology which has impacted widely throughout the chemical industry has been the adoption of solid supported reagents and scavengers as effectors of chemical change and subsequent reaction purification (Ley et al. 2005; Baxendale et al. 2003; Kirschning et al. 2001; Bhattacharyya 2000, 2001; Sherrington 2001; Ley et al. 2000; Thompson 2000; Shuttleworth et al. 2000; Booth and Hodges 1999; Parlow et al. 1999; Drewry et al. 1999; Hodges 1999; Booth and Hodges 1997; Shuttleworth et al. 1997; Kaldor et al. 1996; Chakrabarti and Sharma 1993; Akelah and Sherrington 1981, 1983). However, the advantages provided by utilising these immobilised species have only been partially realised through their application in batch-based chemistries, additional attainable gains can be derived when these agents are incorporated into flow-based synthetic applications.

A few general comments concerning the nature of these immobilised species and their potential application in synthesis are probably pertinent at this stage. Firstly, artificially immobilising a reagent in order to conduct chemical syntheses is not a new concept, having been first reported in 1946 (Sussman 1946). However, tremendous advances have been made in this field, and our group in particular has sought to develop the application of immobilised systems to complex synthetic challenges such as multi-step integrated synthetic operations (Baxendale and Ley

2005; Ley et al. 2001, 2002a, 2002b; Ley and Baxendale 2002a,b; Baxendale et al. 2002; Ley and Massi 2000; Caldarelli et al. 1999a, b, 2000; Habermann et al. 1999a; Habermann et al. 1999b; Hinzen and Ley 1998). Such a philosophy based on continuous sequential or parallel processing of reactants often involving several stages of reactive intermediates has naturally led us to an investigation of such transformations in terms of dynamic coupled multi-step continuous flow procedures (Baxendale et al. 2006a,b,c,d; Saaby et al. 2005).

2 Solid-Supported Reagents

In general terms, *supported reagents* are reactive species that are associated with a support material and are used to transform a substrate (or substrates) into a new chemical product (or products). A key advantage is that it is possible to use an excess of the reagent to drive a reaction to completion and then as the work-up is simplified by the reagent immobilisation, the product can be isolated cleanly (Fig. 1). Additionally, in situations where highly toxic or obnoxious materials must be used, by virtue of the attachment process, such unpleasant compounds can be rendered easier to manipulate and recover, and therefore safer. Likewise, the immobilisation of expensive catalysts and ligands greatly facilitates their recycling potential. Furthermore, immobilisation intrinsically results in reagents becoming site-isolated, that is the immobilised reagents are able to react with substrates in a solution but not readily with one another. This means multi-reagent processes such as oxidation and reduction or other mutually incompatible transformations can be conducted simultaneously in single reaction vessels. This is an ex-

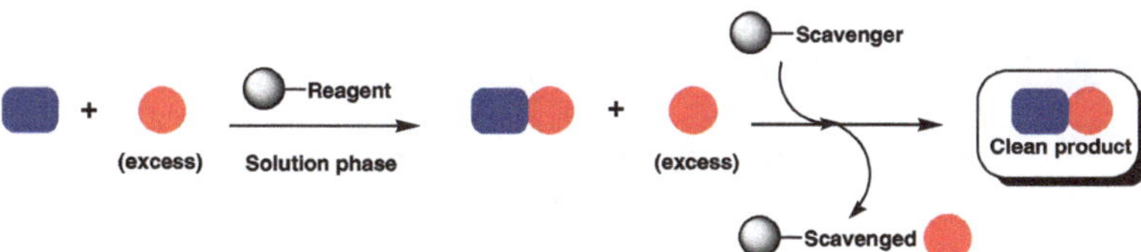

Fig. 1. Basic concept of solid supported reagents

tremely attractive concept and constitutes a significant shift in the way we perceive reactions or plan a chemical synthesis.

However, not all reactions give entirely clean products, but by using supported reagents, by-products and unwanted components can also be scavenged, which obviates the need for more conventional and often more expensive or wasteful procedures such as chromatography, distillation or crystallisation. Purification can also be achieved through the addition of solid-bound species that are programmed to only recognise the target product. This of course is not a serendipitous process but requires specific chemical design. Products can then be captured and impurities washed away before being re-released to give clean compounds, a procedure otherwise known as the catch-and-release technique.

Effective syntheses, convergent as well as batch-splitting, are possible using these methods, which can provide both combinatorial arrays and fragment sets of new molecules (Fig. 2). The act of immobilisation can be accomplished in a variety of ways such as on beads, active

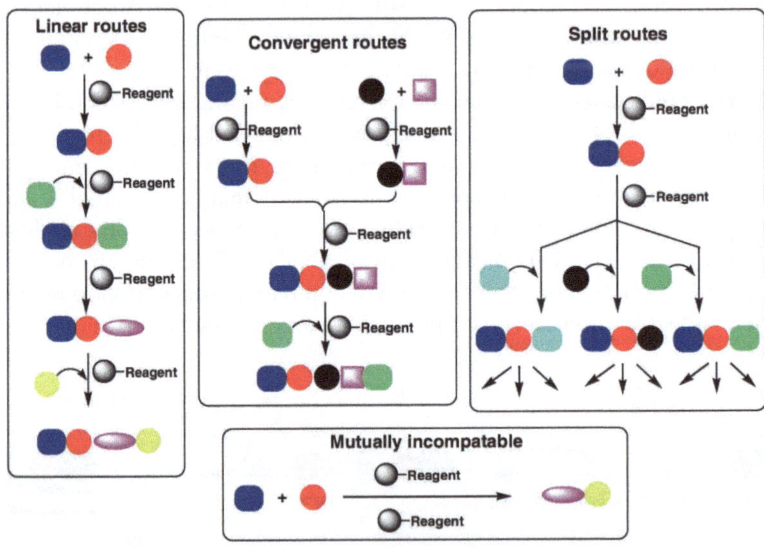

Fig. 2. Application of solid supported reagents

surfaces, colloids, dendrimers, plugs, and laminates. For each of these formats, there is also a variety of support materials available, e.g. polymers, cellulose and silica gels enabling a large selection of reagents to be built.

Utilising these supported reagent procedures has permitted the syntheses of many complex natural products (Ley et al. 2006) to be completed, including the epothilones A and C (Storer et al. 2003, 2004) carpanone (Baxendale et al. 2001, 2002b) and various members of the amaryllidaceae alkaloids family, namely oxomaritidine, epimaritidine (Ley et al. 1999) and plicamine (Baxendale and Ley 2005b; Baxendale et al. 2002c,d) (Fig. 3). In addition, it has been possible to devise protocols to prepare specific drug molecules as well as generate derivatives and analogues of a diverse range of these pharmaceutically interesting compounds (Baxendale and Ley 2005c; Baxendale et al. 2002e,f, 2005; Siu et al. 2004; Baxendale and Ley 2000).

As an example of the sophistication that can be achieved using solid-supported reagents in *multi-step* applications (+)-plicamine (Baxendale and Ley 2005b; Baxendale et al. 2002b,c) truly stands out (Scheme 1). In this synthesis, extensive use was made of parallel optimisation methods in order to progress the forward route more rapidly. The incorporation of focussed microwave techniques also aided in achieving faster and cleaner reactions. Indeed, the whole route, including optimisation, was completed in just 6 weeks, without resorting to any form of traditional separation or purification.

When conducting such syntheses, the immobilised reagents can be delivered in a variety of formats, most commonly their physical form is as loose beads, or as beads contained within pouches or formed into composite plugs. Unfortunately, the commonly used micro-bead reagent configuration is not always the most practical of materials to work with. Measuring and dispensing operations are often associated with problems of static charge and calibration of the stoichiometry. Additional difficulties can also be encountered during the reactions with possible fragmentation and grinding of the beads, which can complicate the work-up protocol. Another issue often highlighted is one of product recovery, which may involve extensive washing of the support in order to fully extract the desired molecules. However, the same reagents can be readily packed into tubes, cartridges and various flow assemblies re-

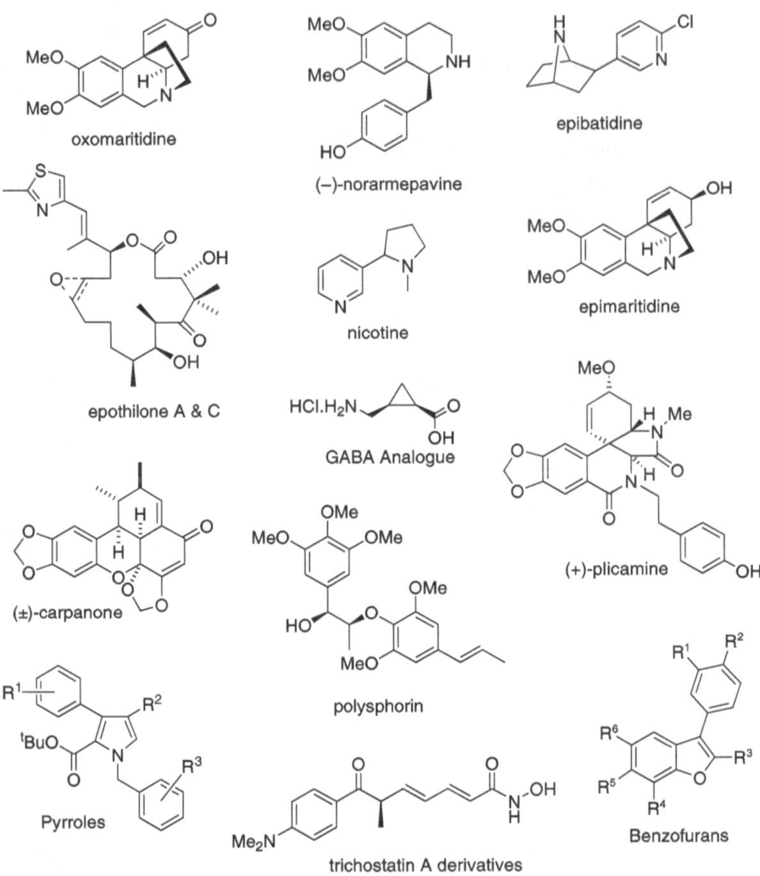

Fig. 3. Natural products and derivatives accessed through solid supported reagent technology

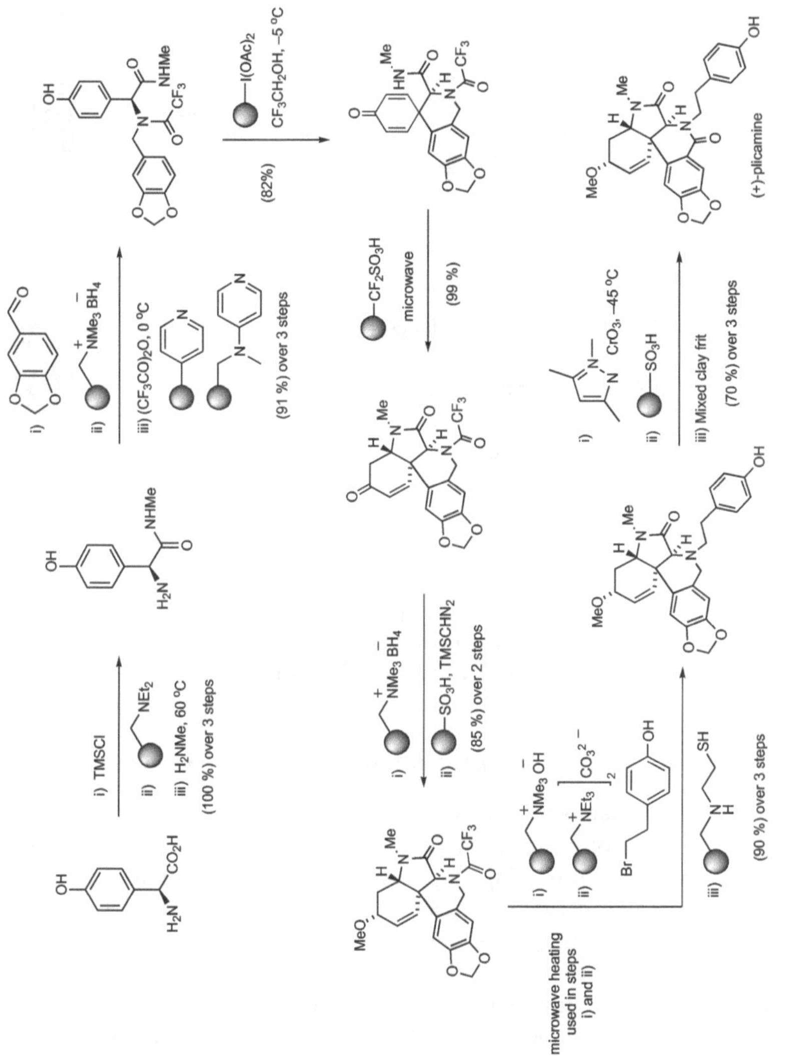

Scheme 1. Synthesis of the natural product (+)-plicamine

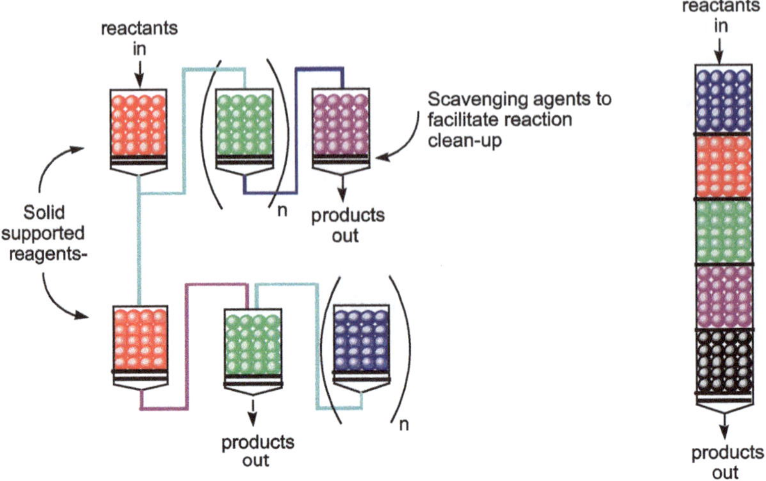

Fig. 4. Assembly of columns using solid supported reagents

sulting in tremendous versatility; particularly so for complex molecule assembly where rapid reaction and reagent optimisation is essential for efficient and high yielding production (Fig. 4).

3 Flow-Based Integration of Supported Reagents

Flow-based chemistry in this context provides the chemist with a new synthesis platform that expands the current capabilities and opportunities available, providing improved safety considerations, enhanced dispensing and mixing coefficients, and real-time diagnostics with the ability to make changes on the fly in a dynamic fashion. As a technique, it replaces traditional glassware with preloaded columns and cartridges containing immobilised reagents and catalysts, also incorporating precision-manufactured reaction chips that permit controlled mixing and precise temperature control of reaction sequences. This essentially transforms conventional batch reaction sequencing into a flowing, dynamic processing procedure by either passing the starting materials through various immobilised reagents or combining intermediates and reagents in specialised reactor blocks in predefined combinations. In this way, the intermediate material generated *in situ* can be subjected to a series of reaction cascades and scavenging protocols before it eventually exits from the reactor into a chemically inactive environment allowing for its immediate collection as a pure product. Throughout this processing, the packed cartridges can be interacted upon by various physical means such as heating/cooling, oscillation, ultrasound, microwaves or irradiation.

Flow chemistry (Luckarift et al. 2005; Doku et al. 2005; Kirschning and Jas 2004; Jönsson et al. 2004; Jas and Kirschning 2003; Watts and Haswell 2003a, 2003b; Fletcher et al. 2002; Hafez et al. 2000, 2002; Hafez 2001; Anderson 2001; Sands et al. 2001; Haswell et al. 2001; Ehrfeld et al. 2000) can also facilitate processes that are not feasible (or at least very technically challenging) by conventional methods, such as the generation and concurrent use of potentially highly reactive or unstable intermediates. In addition, the incorporation of automated and real-time analysis facilitates rapid optimisation and excellent quality control whilst also ensuring reproducibility. By accurate control over flow rates, reactant concentrations, temperatures and pressures, these information-rich sequences generate opportunities for highly responsive modification of chemoselective processes and constitute a paradigm shift for delivery of new chemical entities.

Existing supported reagent technology integrates well into such a synthesis foundation. The extensive range of reagent formats from traditional silica or polymer microbeads through to woven fibers, pressed disks and monolithic structures can all be rapidly customised for use in such flow reactor designs. The development of a diverse tool kit of such functionalised materials can then be combined to facilitate multi-step synthetic sequences. For example, used in series the reaction cartridges could be arranged such that a multitude of reactions could be carried out on each substrate (a splitting of the reaction flow approach), possibly with further monomer inputs at different points allowing access to multiple products (Fig. 5). Alternatively, when used in parallel the reactor arrays could be sequenced for the production and purification of very large libraries of compounds. Clearly, this is an area that is ideally suited to supported reagents, as synthesis in this format is hard to envisage using any other technique since purification on this scale,

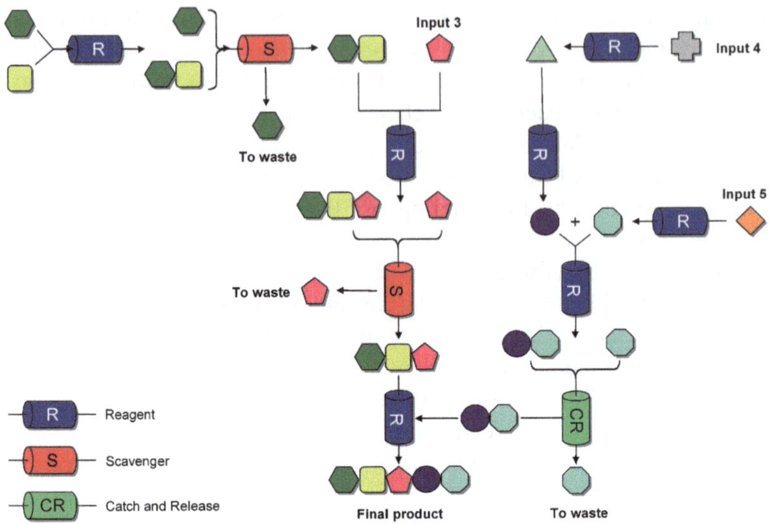

Fig. 5. Flow chemistry principles

whilst not impossible, would make parallelisation prohibitively expensive.

Another value-added consequence of working in a flow mode is the extensive information that can be harvested from the reaction as attained from real-time in-line analysis. Such prospecting can yield extensive information regarding the reaction parameters, allowing rapid optimisation as well as important kinetic and physical chemical data to be archived. Such sampling techniques can be both passive and semi-invasive in nature. Rapid although limited qualitative spectral data can be easily acquired through adjustable wavelength photodiode detectors (or similar spectrometers), which can be assembled as in-line analysis cells or by using the appropriate construction materials can take readings directly through the reactor walls (or connective tubing). Other diagnostic devices can also report on reaction progress using techniques such as impedance measurements, Raman spectroscopy, near or React IR, fluorescence measurements and various bioassays. Alternatively, or in addition, automated sampling techniques can be used to divert small but representative aliquots of the reaction media from the main stream into auxiliary monitoring equipment. This allows powerful techniques such as LC-MS or GC-MS to be assimilated into the system, providing another level of diagnostic capability.

Fundamentally, as each stage of the reaction can be considered spatially isolated from the material proceeding and following it—segmented flow analysis—altering the conditions at any point in the continuum will ultimately allow a contrast at later stages of the reaction's pathway. Thus it becomes a simple task to automatically cycle through various reaction parameters in order to evaluate such effects on the reaction composition and purity. However, a cautionary note should be made regarding how this information is ultimately measured and interpreted depending on whether the reactor is working in a plugged or steady-state operation. Although sequential plugged reactions can exhibit similar characteristics to those that achieve steady state, the material progressing though the system obviously does not experience the same conditions in terms of reagent concentrations or diffusion broadening.

Acquiring this portfolio of reaction information is only the primary stage, effective information feedback and self-optimisation processes using design of experiment (DoE) and principal component analysis (PCA) are highly effective aids (Vickerstaff et al. 2004; Jamieson et al. 2000, 2002). Indeed, the large quantities of data that are captured from the monitoring processes can be directly analysed by such evaluation software, which can in turn be used to directly affect the subsequent operation of the reactor. Not only can this be used to rapidly and efficiently optimise a chemical transformation, but in combination with a flexible reactor design it can also be used to screen for catalyst or even discover new chemistries. Again, it should be emphasised that all this can be done in a fully automated fashion requiring only minimum human intervention following the reactor loading and configuration. It is important that these new technologies should not dominate nor distance the chemist from the synthesis; rather they should be utilised to liberate the chemist from the more mundane and repetitive operations allowing them to reallocate time to the more important tasks of thinking and planning.

4 Flow-Based Chemical Synthesis

Focussed microwave irradiation has always offered an effective mechanism for rapidly heating flow reactions, especially when employing heterogeneous components (Wison et al. 2004; Thomas and Faucher 2000; Kabza et al. 2000; Toteja et al. 1997; Zu and Chuang 1996). The precise control of energy that can be transposed to the sample as well as the highly adjustable temperature profile enables more judicious heating in the face of constantly changing flow rates and reactant concentrations. In a recent study of non-metal-catalysed intramolecular cyclotrimerisation reactions we demonstrated it was possible to conduct such chemistry as a flow process at elevated temperatures using a glass coil inserted into the microwave reaction cavity (Fig. 6) (Saaby et al. 2005). Gram quantities of material could thus be prepared by pumping a DMF solution of the precursor through the system heated at 200°C whilst maintaining reaction pressure by using a back pressure regulator. Interestingly, in the corresponding sealed batch system the decomposition of DMF used as the solvent led to the generation of significant pressures. Using a continuous flow process avoided this problem by restricting the residence time of the DMF and constantly balancing the internal pressure, thereby reducing the decomposition and any pressure increase.

Microwaves have also proved useful for the preparation of various substituted pyrazole structures that are prepared in flow by using a modified tubing reactor inserted into the microwave well (Fig. 7). Several meters (10–15 m) of microwave transparent tubing could be tightly spiralled around a Teflon column manufactured to fit into the microwave and mimic the action of a normal batch microwave vial. Connecting the input of the reactor to a reactant reservoir via a HPLC pump and the output to a twin column assembly containing a polymeric primary amine scavenger and activated carbon decolouriser completed the system. This simple flow setup has been used for periods of 36 h at temperatures up to 150°C in order to prepare 15 g batches of bulk intermediates in high yields and excellent purities (Fig. 8).

One of the obvious areas where flow chemistry comes into its own is in the arena of catalysed reactions and continuous processing. An excellent example of this is the Suzuki reaction (for recent reviews on the Suzuki Cross-Coupling reaction, see Miura 2004; Bellina et al. 2004;

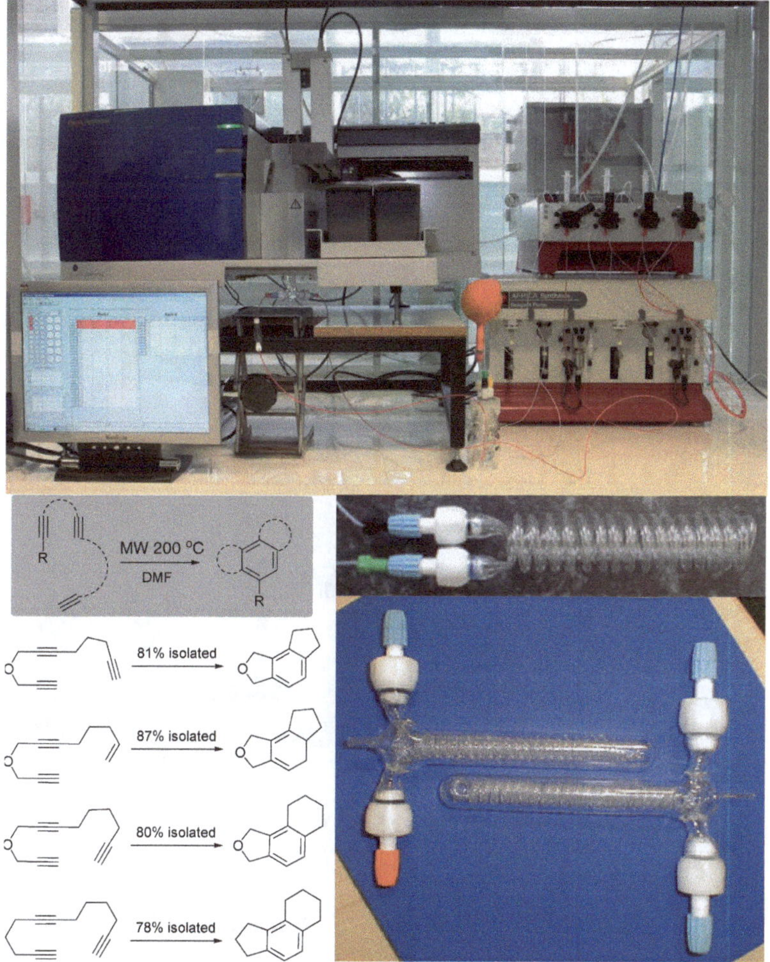

Fig. 6. Intramolecular cyclotrimerisation reactions, equipment and glass microwave insert reactors

Pershichini 2003; Kotha et al. 2002), which is by far the most versatile synthetic method available for the generation of unsymmetrical biaryl compounds and so was an obvious target for the development of flow

Fig. 7. Insert microwave tubing reactor

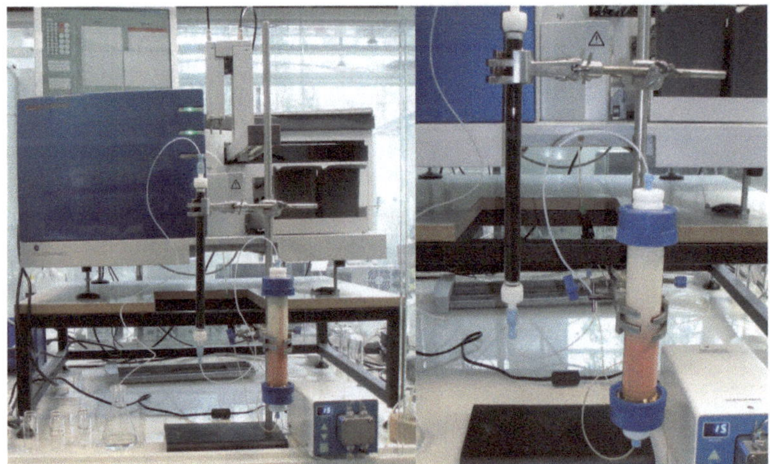

Fig. 8. Column attachments for microwave cavity tubing reactor

reactor designs. The standard catalyst employed in this reaction is a palladium species; however, the reactions suffer from high levels of residual palladium contamination of the final products. We have previously reported a polyurea microencapsulated palladium catalyst (Pd EnCat™;

Fig. 9), which has been applied to a range of cross-coupling and reduction reactions (Pd EnCat™ 30 available from Aldrich 644714-XXXG [XXX = 1, 10, 100 with regard to batch size] loading: 0.4 mmol/g Pd; Yu et al. 2003; Ramarao et al. 2002; Ley et al. 2002; Bremeyer et al. 2002). This type of polyurea matrix is able to ligate and retain the palladium within the cavities of the microcapsule but permits the reactants to diffuse to the active catalyst and then allows the products to exit back into solution. Indeed, inductively coupled plasma (ICP) analysis of product mixtures has proven palladium levels to be less than 10 ppm corresponding to less than 1% leaching of the original palladium content of the capsules. Additionally, should leaching become a problem, further passage through a Quadrapure™ TU scavenging column leads to enhanced quality products. The Pd EnCat catalyst has been used in two sets of Suzuki coupling reactions in flow under very different reaction conditions.

In a proof of concept study, it was shown that iodobenzene and *p*-toylboronic acid could be quantitatively coupled at 70°C when passed through a fixed bed of the Pd EnCat™ in the presence of an equivalent of the $^{n}Bu_4NOMe$ salt as a 9:1 Toluene/MeOH solution (Lee et al. 2005). The reaction was complete following only a single pass through the loaded HPLC column (5 × 0.45 cm ID) using a flow rate of 0.2 ml/min resulting in a mean residence time of approximately 4 min. In addition, the catalyst system could be recycled multiple times without any noticeable deterioration of the catalyst system. Such an example therefore aptly demonstrates the potential for process intensification that such a combination of novel technologies represent.

Following on from this work, it was further demonstrated that microwave heating could be used to vastly expedite the production of Suzuki furnished compound arrays including the generation of multigram quantities of products in a single, continuous operation (Baxendale et al. 2006b). The reactor design used was based upon a simple continuous glass U-tube which was packed with the heterogeneous catalyst and inserted into the microwave cavity (Fig. 10). Stock solutions of the boronic acid, aryl halide and $^{n}Bu_4NOAc$ activator were prepared and constantly fed at a flow rate of 0.1 ml/min through the flow reactor that was resident in the microwave cavity (Fig. 11). This configuration gave a total residence time of 225 s for the reactor assembly, although

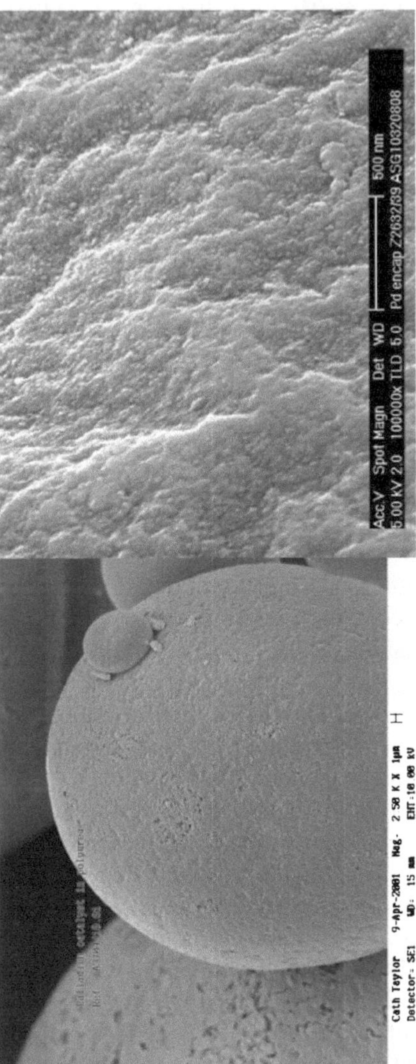

Fig. 9. Magnified Pd EnCat catalysts bead and high-resolution electron microscope image of the surface of the polyurea matrix

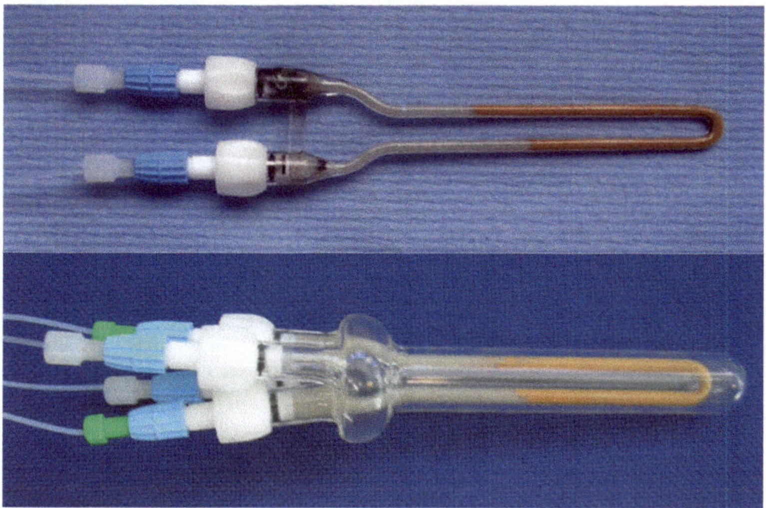

Fig. 10. Pd EnCat microwave reactor columns, simple U-tube and enclosed system

for only 65 s of this time was the solution present in the microwave cavity. As the reaction mixture exited the reactor chamber, it was progressed through a column of Amberlyst 15 sulfonic acid resin to remove any residual base and boronic acid salts. The solution could then be collected, and following evaporation the final product isolated without the need for further purification.

Interestingly, applying a cooling stream of compressed gas to the U-tube reactor while maintaining the microwave irradiation was found to further enhanced the yield and purity of the process. Indeed, microwave heating has been shown to significantly improve the yields of reactions in which the reactants or products suffer from thermal decomposition because of the reduced reaction times and the more even application of energy (for general reviews on the use of microwave irradiation in organic synthesis, see Kappe 2004; Nüchter et al. 2004; Wathey et al. 2002; Wathey 2001). As an extension of this effect, techniques involving microwave irradiation in heat/cool cycles or pulsed power sequences used to limit the global temperature as well as using

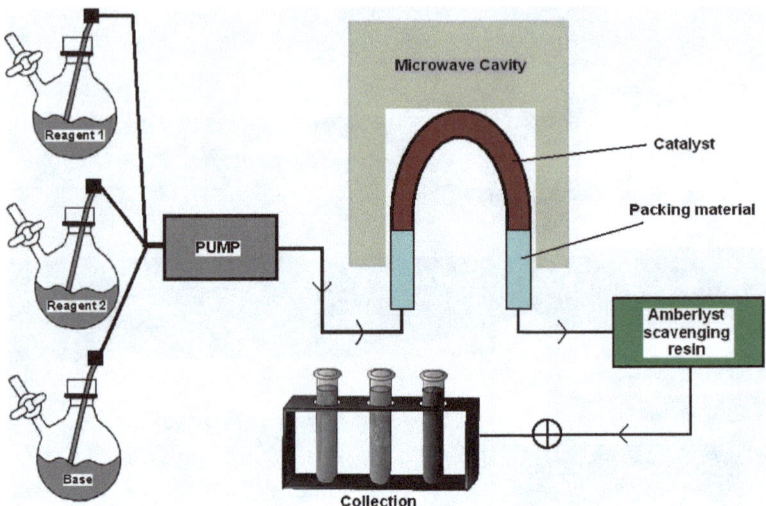

Fig. 11. Reactor configuration for flow Suzuki reactions

external cooling have all generated products of significantly higher purity (Arvela and Leadbeater 2005; Leadbeater et al. 2005; Humphrey et al. 2003; Chen and Deshpande 2003).

In theory immobilisation of the catalyst in such a packed column format permits an easy automated recycling process to be established. To validate this notion, the complete set of ten coupling reactions outlined in Table 1 were performed. Each substrate pairing was sequentially passed through the same reactor column without regeneration or replacement of the catalyst. The individual streams of substrates were separated using a blank solvent plug which also functioned to wash the catalyst bed before the next set of reactants arrived. Following this protocol, no cross-contamination of the products was detected. Clearly such a processing procedure is of intrinsic value to the high-throughput automated synthesis of compound libraries.

In addition to the ability to prepare multiple products, it was also demonstrated that it was possible to scale-out the reaction using prolonged reaction times. In this way, multi-gram quantities of material could be readily isolated (Scheme 2). The reactions were very efficient

Entry	Boronic Acid	Halide	Yield% (Purity%)
1	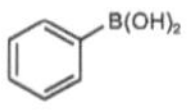 $B(OH)_2$	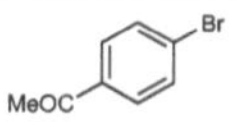 Br, MeOC	90 (>98)
2	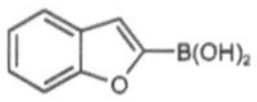 $B(OH)_2$, O	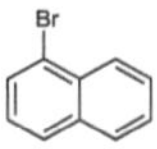 Br	87 (>98)
3	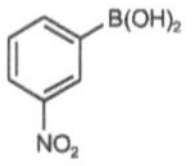 $B(OH)_2$, NO_2	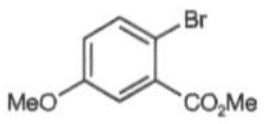 Br, MeO, CO_2Me	94 (>98)
4	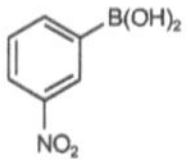 $B(OH)_2$, NO_2	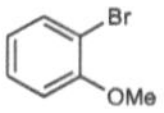 Br, OMe	92 (92)
5	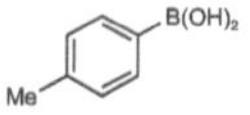 $B(OH)_2$, Me	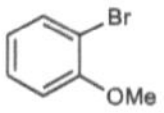 Br, OMe	88 (94)
6	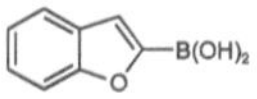 $B(OH)_2$, O	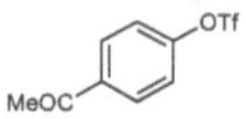 OTf, MeOC	97 (91)
7	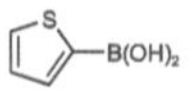 S, $B(OH)_2$	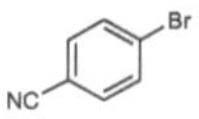 Br, NC	76 (>98)
8	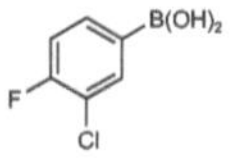 $B(OH)_2$, F, Cl	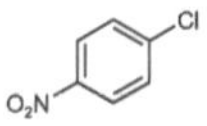 Cl, O_2N	89 (>98)
9	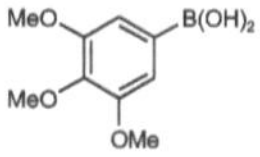 MeO, $B(OH)_2$, MeO, OMe	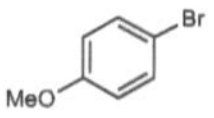 Br, MeO	94 (91)
10	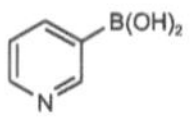 $B(OH)_2$, N	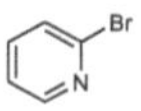 Br, N	95 (82)

Table 1. Sequential processing of multiple substrates in flow [1 equiv. halide, 1 equiv. boronic acid, 2 equiv. Bu_4NOAc in EtOH. Flow rate 0.1 mL/min, 50 W with external cooling]

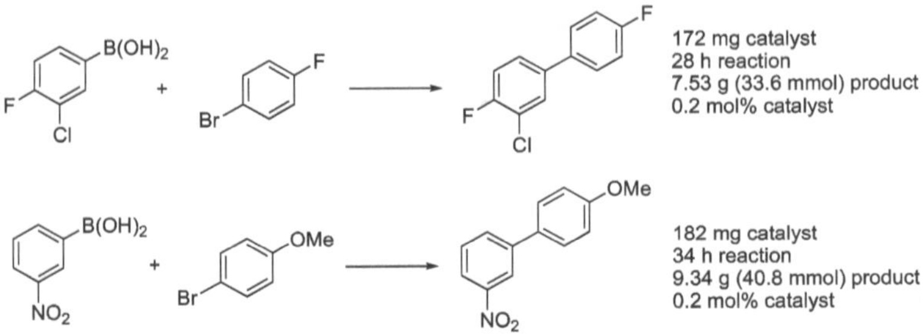

172 mg catalyst
28 h reaction
7.53 g (33.6 mmol) product
0.2 mol% catalyst

182 mg catalyst
34 h reaction
9.34 g (40.8 mmol) product
0.2 mol% catalyst

Scheme 2. Scaled synthesis of Suzuki reactions

progressing for several hours, after which the conversion dropped dramatically, finally deactivating the catalyst. In this process, the yields of products isolated correspond to an ultimate catalyst loading of as little as 0.2 mol%, which is exceptionally low for this type of catalyst. The high catalytic turnover and short residence times required are certainly a consequence of the high effective catalyst-to-substrate ratios encountered within the reactor. What emerges of considerable importance from this study is the demonstration that a single reactor design can be utilised both for library generation and scaled synthesis.

Additional advantages in terms of substrate selectivity were also observed from the extremely short reaction times and mild reaction conditions. For example, the reactive 2-iodobromobenzene was used as a substrate because in the corresponding batch reaction with 2-benzofuran boronic acid, large quantities of the double addition product were detected (Scheme 3). However, when the same coupling was performed in flow, the reaction could be optimised so that boronic acid reacted only

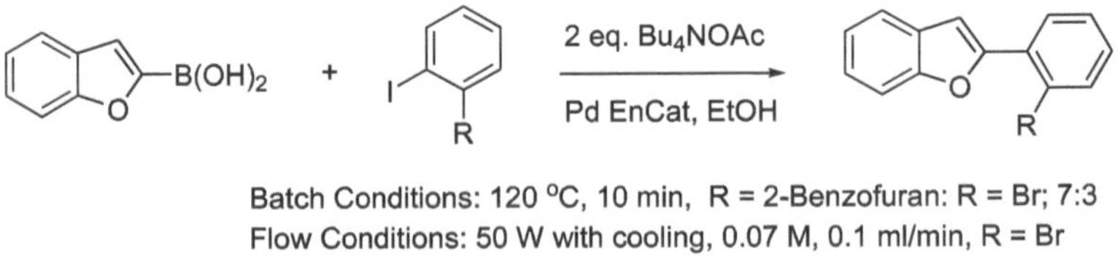

2 eq. Bu_4NOAc
Pd EnCat, EtOH

Batch Conditions: 120 °C, 10 min, R = 2-Benzofuran: R = Br; 7:3
Flow Conditions: 50 W with cooling, 0.07 M, 0.1 ml/min, R = Br

Scheme 3. Enhanced selectivity in flow microwave reactions at the iodide site, generating the single substitution product. This example of enhanced selectivity clearly demonstrates the additional synthetic scope that can be provided from working in a flow domain. This is particularly important when one considers the extended range of substrates that could be employed and the additional reaction sequences leading to increased molecular diversity that could be achieved.

We have also employed conventional conduction heating techniques using modified heating blocks for both micro-channel reactor chips and packed glass columns (Fig. 12). In addition, our need for specific temperature control, which may differ for each stage of the reaction sequence, has been achieved using a dedicated column heater the Vapourtec R-4 (Fig. 12). Such equipment was used to great effect in the recent synthesis of a collection of 4,5-disubstituted oxazoles (Baumann et al. 2006). The reaction scheme involved the additional reaction of an isonitrile moiety to an acyl chloride followed by a base catalysed intramolecular cyclisation (Scheme 4). The chemistry was successfully optimised and the compounds repeatedly prepared on-demand using a bespoke automated modular flow reactor as described in Fig. 13. In total, 37 different compounds were synthesised in excellent yields (Fig. 14) in amounts ranging from 150 mg to 10 g of material. Apart from the introduction of a scavenging cartridge of Quadrapure™ BZA, a primary amine functionalised macroporous polymer, no further purification was required in order to isolate pure materials. The ability to carry out such chemical reactions at various scales in a controlled automated fashion further exemplifies the power of such processing procedures.

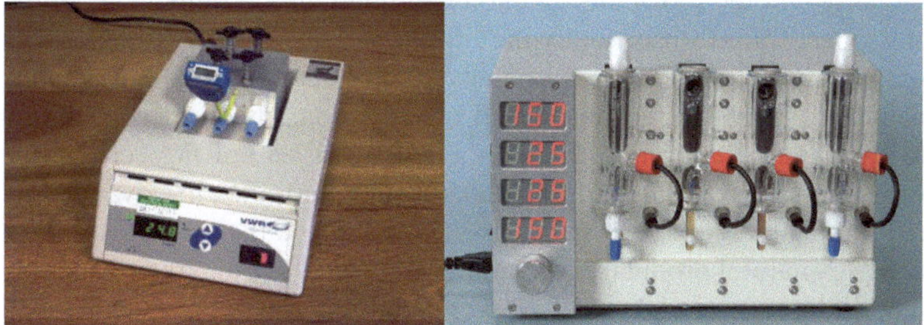

Fig. 12. Column heating block and Vapourtec R-4 Flow Reactor Heater

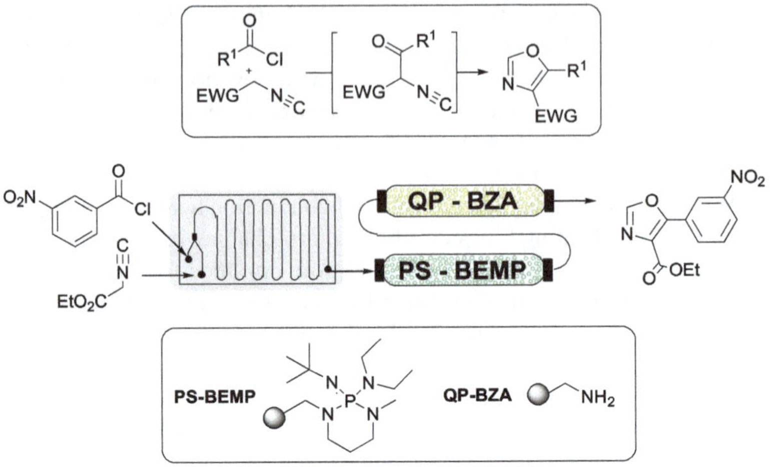

Scheme 4. Preparation of 4,5-disubstituted oxazoles

5 Natural Product Synthesis

So far throughout this article, a key principle has been the on-demand manufacture of advanced chemical materials from integrated *multi-step* sequences. Obviously, single-step chemical transformations have their place, and a rapidly growing literature base of such operations can also be uncovered in the recent chemical literature. However, the ultimate power of synthesis is in its ability to construct architecturally and individually tailored molecules of choice through multi-component, *multi-step* operations. Consequently the adoption of flow-based synthesis will only really be truly accepted when it can supersede existing batch-based chemical processing techniques. It has therefore always been our aim to focus our attention on the more challenging end of flow chemistry investigations by evaluating extended synthetic routes to advanced final products. In this respect, we have investigated the synthesis of two complex natural products, grossamide (Baxendale et al. 2006a) and oxomaritidine (Baxendale et al. 2006c) as initial targets.

The synthesis of grossamide (Baxendale et al. 2006a) is interesting because it brings together a number of concepts previously men-

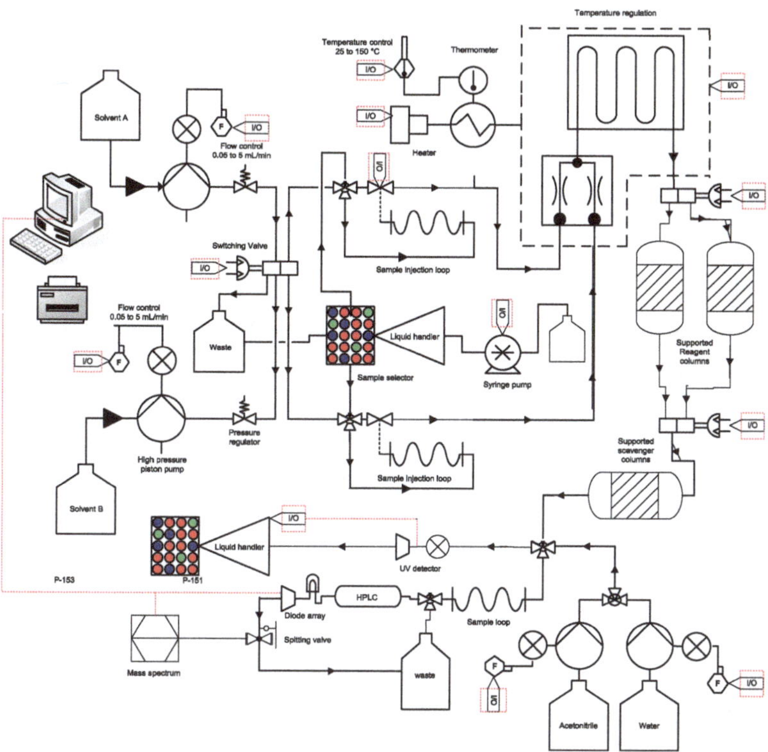

Fig. 13. Schematic of microflow reactor system

tioned and also introduces the practice of using immobilised enzymes under continuous flow conditions. Although the overall complexity of this neolignan may indicate on first inspection a longer sequence of steps, in fact this compound was prepared via only two transformations (Scheme 5). By eluting a prepacked column of immobilised HOBt with a solution of ferulic acid and activator, the corresponding tethered activated ester could be formed. This allows excess starting material and by-products to be washed from the column. This procedure was also used to great effect as an activation step and purification aid in the synthesis of sildenafil (Baxendale and Ley 2000). Next a solution containing an

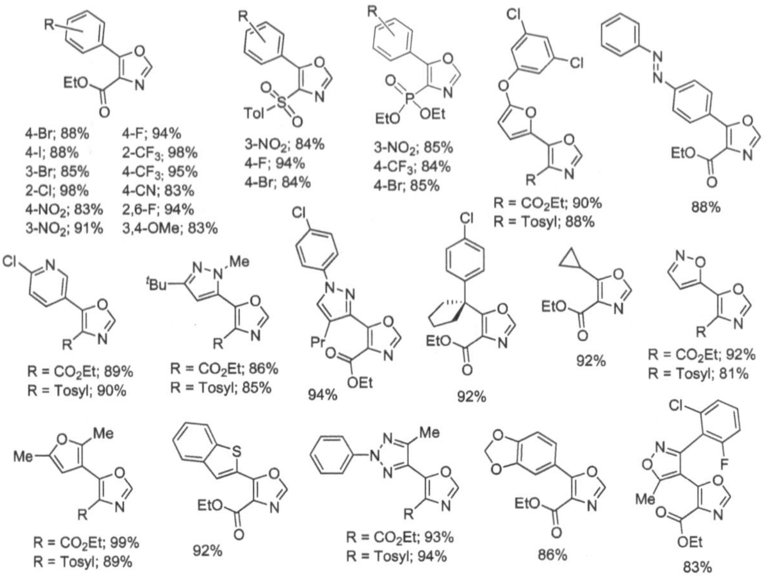

Fig. 14. Oxazoles prepared using flow reaction conditions

amine-coupling agent was directed through the loaded column resulting in the formation of the coupled amide product. As these two sequences are distinct in nature, two parallel columns can be run in conjunction so that as one is loading to form the active ester the previously functionalised column is undergoing reaction to generate the desired product. Obviously, the two columns can be rapidly interconverted, thereby giving in essence a seamless continuous feed of the product. Directing this flow stream through a column of supported sulfonic acid removed any excess unreacted amine residues before it was mixed with a second solution containing hydrogen peroxide and a buffer. The combined reagent stream was then passed through a final column containing horseradish peroxidase enzyme which was trapped on a modified silica support. The resulting solution could then be collected and the product isolated. Although this reaction scheme only yielded a single product, it is easy to envisage similar reactions based on entirely unnatural coupling deriva-

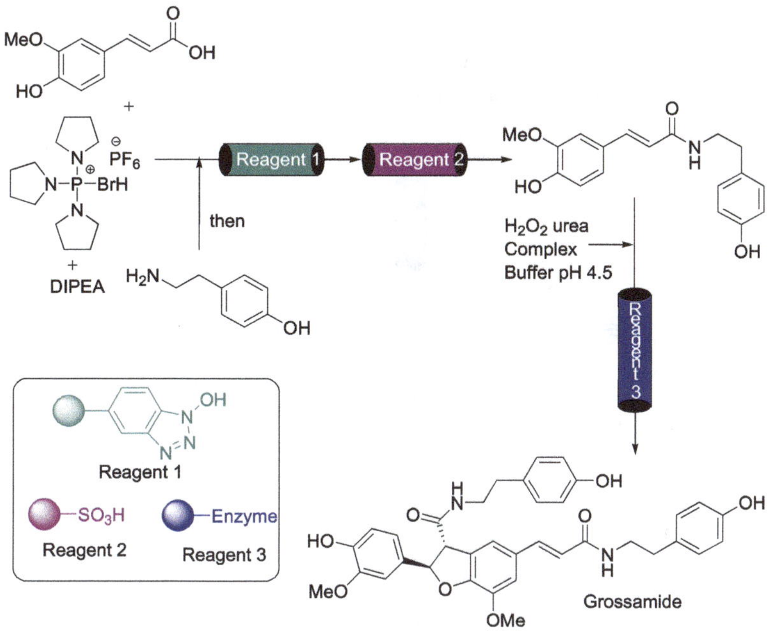

Scheme 5. Flow synthesis of the natural product Grossamide

tives or the generation of extended routes yielding more elaborate structures.

More advanced peptides have also been prepared in a similar automated flow sequence through sequential activation of the carboxylic motif in a linear fashion (Baxendale et al. 2006d). Initially a series of the protected dipeptides were generated (Table 2), but this process has now been extended to the formation of tripeptides (Scheme 6) and could easily be further modified to allow additional couplings.

Our most significant challenge to date has been the synthesis of oxomaritidine as a flow-derived process. The Amaryllidaceae alkaloid oxomaritidine had been previously synthesised using both traditional solution-phase methodology (Schwartz and Holton 1970) and with the aid of polymer-supported reagents (Ley et al. 1999), making it an excellent benchmark for our evaluation. Indeed the time savings resulting from

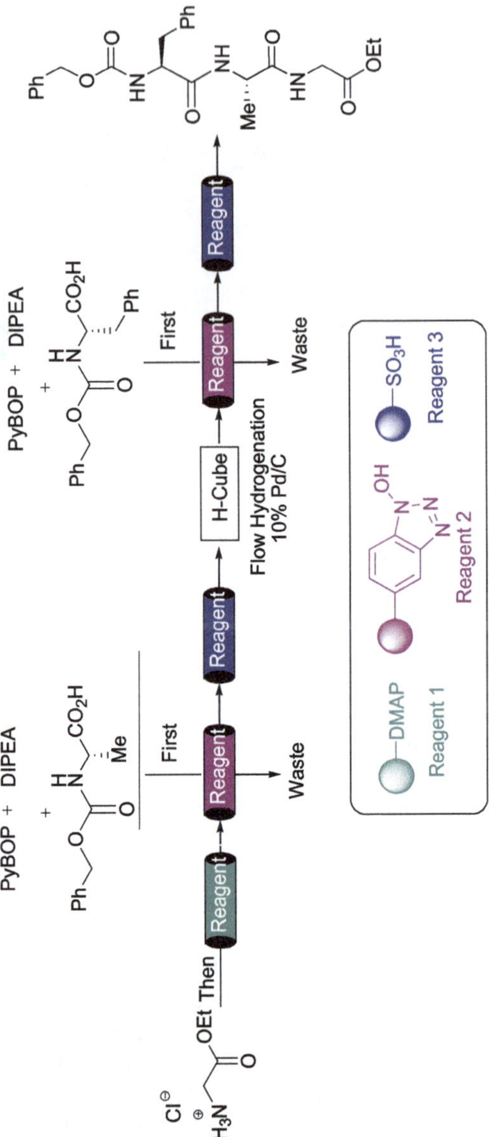

Scheme 6. Sequence for the construction of tripetides

Table 2 Dipeptide structures prepared

Entry	Protecting group	R	R_1	R_2	Isolated yield
1	Boc	Me	CH_2Ph	Et	80%
2	Boc	Me	H	Et	81%
3	Boc	Me	$CHMe_2$	Me	83%
4	Boc	Me	$–(CH_2)_3–$	Me	66%
5	Cbz	CH_2Ph	$CHMe_2$	Me	79%
6	Cbz	Ch_2Ph	H	Et	76%
7	Cbz	Me	CH_2Ph	Et	75%
8	Cbz	Me	H	Et	78%
9	Cbz	Me	$–(CH_2)_3–$	Me	61%

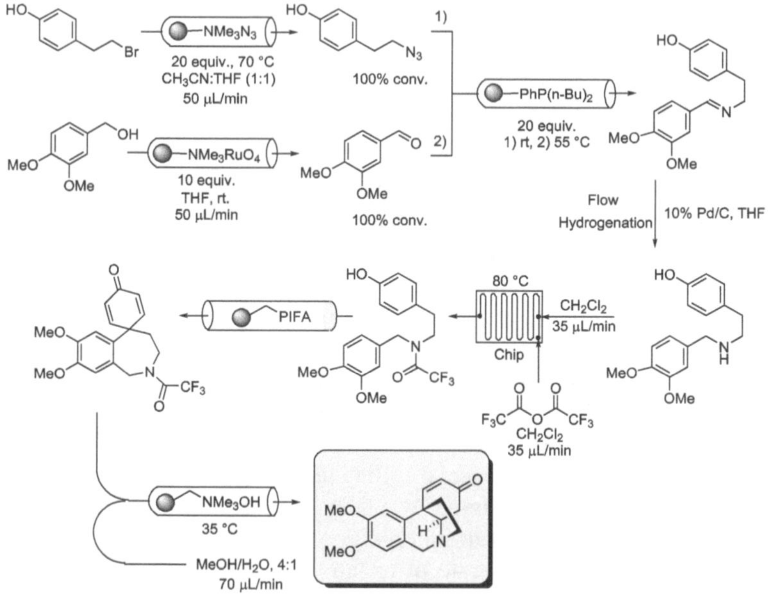

Scheme 7. Continuous flow synthesis of the natural product Oxomaritidine

the application of these flow-through methods in comparison to conventional procedures are dramatic. The synthesis time for the flow-through process to produce the natural product in an automated sequence from readily available starting materials was less than a day (Scheme 7). Originally, this synthesis had taken approximately 4 working days, including quenching procedures and column chromatography to purify the individual products at various stages. By streamlining the purification requirements using solid-supported reagent technology, it was possible to reduce this timeframe to around 2 days. Obviously, the use of the supported reagents grants a considerable productivity savings, although such procedures still involve a significant level of human intervention, as required by reaction monitoring and the filtration/washing steps. It should also be highlighted that although a certain degree of setting up time is required in the flow-through process, once the system is configured the equipment can function independently of the human operator. Therefore, it is easy to see why such protocols could revolutionise the working practices within most laboratories.

6 Conclusion

Commercial synthesis requirements are rapidly changing, generating an inherent need to conduct chemistry in a more efficient and timely manner. In addition, as the global emphasis towards more eco-friendly and sustainable practices unfolds before us, we are being required to re-evaluate how chemical synthesis should be conducted. We are already applying the principles of this new paradigm to environmentally cleaner and more efficient chemical processes, but more progress is required. It is our belief that a new approach to organic synthesis will have to be embraced and that this will not come from small step changes but will require a major adjustment in strategic approach. The aspects of chemical synthesis, physiochemical profiling and bio-evaluation will need to become more closely coupled, whether for pharmaceutical discovery purposes or new material design the arguments are the same. Synthesis procedures cannot operate in isolation to their intended products, such an approach leads to wasted synthesis time and the generation of unwanted material. A more integrated and continuous relay of information regarding the ongoing synthesis and its products in terms of basic characterisation, physical properties and their functions needs to be captured. Immediate collection and analysis of such data will enable more educated and responsive choices to be made, varying from mundane factors such as reaction optimisation to higher-level decisions about what molecules should be synthesised next. The requirements in terms of information retrieval and flexibility of synthetic implementation seem ideally suited to a flow-based approach to chemical synthesis. It is therefore our conclusion that future synthesis programmes will make increasing use of flow techniques and many of these will benefit from the use of solid-supported reagents and scavengers.

References

Akelah A, Sherrington DC (1981) Application of functionalized polymers in organic synthesis. Chem Rev 81:557–587

Akelah A, Sherrington DC (1983) Recent developments in the application of functionalized polymers in organic synthesis.Polymer 24:1369–1386

Anderson NG (2001) Practical use of continuous processing in developing and scaling up laboratory processes. Org Proc Res Dev 5:613–621

Arvela RK, Leadbeater NE (2005) Suzuki coupling of aryl chlorides with phenylboronic acid in water, using microwave heating with simultaneous cooling. Org Lett 7:2101–2104

Baumann M, Baxendale IR, Ley SV, Smith CD, Tranmer GK (2006) Fully Automated Continuous Flow Synthesis of 4,5-Disubstituted Oxazoles. Org Lett 9:1521–1524

Baxendale IR, Ley SV (2000) Polymer-supported reagents for multi-step organic synthesis: application to the synthesis of sildenafil. Bioorg Med Chem Lett 10:1983–1986

Baxendale IR, Ley SV (2005a) Synthesis of alkoloid natural products using solid supported reagents and scavengers. Curr Org Chem 9:1521–

Baxendale IR, Ley SV (2005b) Synthesis of the alkaloid natural products (+)-plicane and (–)-obliquine, using polymer-supported reagents and scavengers. Ind Eng Chem Res 44:8588–8592

Baxendale IR, Ley SV (2005c) Formation of 4-aminopyrimidines via the trimerization of nitriles using focused microwave heating. J Combinatorial Chem 7:483–489

Baxendale IR, Lee A-L, Ley SV (2001) A concise synthesis of the natural product carpanone using solid-supported reagents and scavengers. Synlett 482–484

Baxendale IR, Ley SV Sneddon H (2002a) A clean conversion of aldehydes to nitriles using a solid-supported hydrazine. Synlett 775–777

Baxendale IR, Lee A-L, Ley SV (2002b) A concise synthesis of carpanone using solid-supported reagents and scavengers. J Chem Soc Perkin Trans 1:1850–1867

Baxendale IR, Ley SV, Nesi M, Piutti C (2002c) Total Synthesis of the amaryllidaceae alkaloid (+)-plicamine and its unnatural enantiomer by using solid-supported reagents and scavengers in a multistepsequence of reactions. Angew Chem Int Ed 41:2194–2197

Baxendale IR, Ley SV, Piutti C, Nesi M (2002d) Total synthesis of the amaryllidaceae alkaloid (+)-plicamine using solid-supported reagents. Tetrahedron 58:6285–6304

Baxendale IR, Ley SV, Ernst M, Krahnert W-R (2002e) Application of polymer-supported enzymes and reagents in the synthesis of γ-aminobutyric acid (GABA) analogues. Synlett 1641–1644

Baxendale IR, Brusotti G, Matsuoka M, Ley SV (2002f) Synthesis of nornicotine, nicotine and other functionalised derivatives using solid-supported reagents and scavengers. J Chem Soc Perkin Trans 1 2:143–154

Baxendale IR, Storer RI, Ley SV (2003) Supported reagents and scavengers in multi-step organic synthesis in polymeric materials in organic synthesis and catalysis. Ed Buchmeiser MR VCH, Berlin

Baxendale IR, Ley SV, Martinelli M (2005) The rapid preparation of 2-aminosulfonamide-1,3,4-oxadiazoles using polymer-supported reagents and microwave heating. Tetrahedron 61:5323–5349

Baxendale IR, Griffiths-Jones CM, Ley SV, Tranmer GK (2006a) Preparation of the neolignan natural product grossamide by a continuous-flow process. Synlett 427–430

Baxendale IR, Griffiths-Jones CM, Ley SV, Tranmer GK (2006b) Microwave-assisted Suzuki coupling reactions with an encapsulated palladium catalyst for batch and continuous-flow transformations. Chem Eur J 12:4407–4416

Baxendale IR, Deeley J, Griffiths-Jones CM, Ley SV, Saaby S, Tranmer GK (2006c) A flow process for the multi-step synthesis of the alkaloid natural product oxomaritidine: a new paradigm for molecular assembly. J Chem Soc Chem Commun 2566–2568

Baxendale IR, Ley SV, Smith CD, Tranmer GK (2006d) A flow reactor process for the synthesis of peptides utilizing immobilized reagents, scavengers and catch and release protocols. J Chem Soc Chem Commun 4835–4837

Bellina F, Carpita A, Rossi R (2004) Palladium catalysts for the Suzuki Cross-coupling reaction: an overview of recent advances. Synthesis 2419–2440

Bhattacharyya SJ (2000) Polymer-supported reagents and catalysts recent advances in synthetic applications. Comb Chem High Throughput Screening 3:65–92

Bhattacharyya SJ (2004) New developments in polymer-supported reagents, scavengers and catalysts for organic synthesis. Curr Opin Drug Discov Devel 7:752–764

Booth RJ, Hodges JC (1997) Polymer-supported quenching reagents for parallel purification. J Am Chem Soc 19:4882–4886

Booth RJ, Hodges JC (1999) Solid-supported reagent strategies for rapid purification of combinatorial synthesis products. Acc Chem Res 32:18–26

Bremeyer N, Ley SV, Ramarao C, Shirley IM, Smith SC (2002) A flow reactor process for the synthesis of peptides utilizing immobilized reagents, scavengers and catch and release protocols. Synlett 1843–1844

Caldarelli M, Habermann J, Ley SV (1999a) Clean five-step synthesis of an array of 1,2,3,4-tetra-substituted pyrroles using polymer-supported reagents. J Chem Soc., Perkin Trans 1 107–110

Caldarelli M, Habermann J, Ley SV (1999b) Synthesis of an array of potential matrix metalloproteinase inhibitors using a sequence of polymer-supported reagents. Biorg Med Chem Lett 9:2049–2052

Caldarelli M, Baxendale IR, Ley SV (2000) Clean and efficient synthesis of azo dyes using polymer-supported reagents. J Green Chem 43–45

Chakrabarti A, Sharma MM (1993) Cationic ion exchange resins as catalyst. React Polym 20:1–45

Dye JL, Lok MT, Tehan FJ, Creaso JM, Voorhees KJ (1973) Flow synthesis. Substitute for the high-dilution steps in cryptate synthesis. J Org Chem 38:1773–1775

Drewry DH, Coe DM, Poon S (1999) Solid-supported reagents in organic synthesis. Med Res Rev 19:97–148

Doku GN, Verboom W, Reinhoudt DN, van den Berg A (2005) On-microchip multiphase chemistry—a review of microreactor design principles and reagent contacting modes. Tetrahedron 61:2733–2742

Ehrfeld W, Hessel V, Löwe H (2000) Microreactors: new technology for modern chemistry. Wiley-VCH, Weinheim, Germany

Fletcher PDI, Haswell SJ, Pombo-Villar E, Warrington BH, Watts P, Wong SYF, Zhang X (2002) Micro reactors: principles and applications in organic synthesis. Tetrahedron 58:4735–4757

Haberman. Chem Eur J 8:4115–4119

Haswell SJ, Middleton RJ, O'Sullivan B, Skelton V, Watts P, Styring PJ (2001) The application of micro reactors to synthetic chemistry. Chem Soc Chem Commun 5:391–398

Hinzen B, Ley SV (1998) Synthesis of isoxazolidines using polymer supported perruthenate (PSP). J Chem Soc Perkin Trans 11–2

Hodges JC (1999) Covalent scavengers for primary and secondary amines. Synlett 1:152–158

Humphrey CE, Easson MAM, Tierney JP, Turner N (2003) Capture and release reagent for the synthesis of amides and novel scavenger resin. J Org Lett 5:849–852

Jamieson C, Congreave MS, Emiabata-Smith DF, Ley SV (2000) A rapid approach for the optimisation of polymer supported reagents in synthesis. Synlett 603–607

Jamieson C, Congreave MS, Emiabata-Smith DF, Ley SV, Scicinski J (2002) application of ReactArray robotics and design of experiments techniques in optimisation of supported reagent chemistry. J Org Proc Res Dev 6:823–825

Jas G, Kirschning A (2003) Continuous flow techniques in organic synthesis. Chem Eur J 9:5708–5723

Jönsson D, Warrington BH, Ladlow M (2004) Automated flow-through synthesis of heterocyclic thioethers. J Comb Chem 6:584–595

Kabza KG, Chapados BR, Gestwicki JE, McGrath JL (2000) Microwave-induced esterification using heterogeneous acid catalyst in a low dielectric constant medium. J Org Chem 65:1210–1214

Kaldor SW, Siegel MG, Dressman BA, Hahn PJ (1996) Use of solid supported nucleophiles and electrophiles for the purification of non-peptide small molecule libraries. Tetrahedron Lett 37:7193–7196

Kappe CO (2004) Controlled microwave heating in modern organic synthesis. Angew Chem Int Ed 43:6250–6284

Karnatz FA, Whitmore FC (1932) Dehydration of diethylcarbinol. J Am Chem Soc 54:3461

Kirschning A, Jas G (2004) Applications of immobilized catalysts in continuous flow processes. Topics in Curr Chem 209–239

Kirschning A, Monenschein H, Wittenberg R (2001) Functionalized polymers – emerging versatile tools for solution-phase chemistry and automated parallel synthesis. Angew Chem Int Ed 40:650–679

Kotha S, Lahiri S, Kashinath D (2002) Recent applications of the Suzuki–Miyaura cross-coupling reaction in organic synthesis. Tetrahedron 58:9633–9695

Leadbeater NE, Pillsbury SJ, Shanahan E, Williams VA (2005) An assessment of the technique of simultaneous cooling in conjunction with microwave heating for organic synthesis. Tetrahedron 61:3565–3585

Lee CKY, Holmes AB, Ley SV, McConvey IF, Al-Duri B, Leeke GA, Santos RCD, Seville JPK (2005) Efficient batch and continuous flow Suzuki cross-coupling reactions under mild conditions, catalysed by polyurea-encapsulated palladium (II) acetate and tetra-*n*-butylammonium salts. J Chem Soc Chem Commun 2175–2177

Ley SV, Baxendale IR (2002a) Organic synthesis in a changing world. Chem Rec 2:377–388

Ley SV, Baxendale IR (2002b) New tools and concepts for modern organic synthesis. Nat Rev Drug Disc 1:573–586

Ley SV, Massi A (2000) J Comb Chem Polymer supported reagents in synthesis: preparation of bicyclo[2.2.2]octane derivatives via tandem michael addition reactions and subsequent combinatorial decoration. 2:104–107

Ley SV, Schucht O, Thomas AW, Murray PJ (1999) Synthesis of the alkaloids (±)-oxomaritidine and (±)-epimaritidine using an orchestrated multi-step sequence of polymer supported reagents. J Chem Soc Perkin Trans 1 1251–1252

Ley SV, Baxendale IR, Bream RN, Jackson PS, Leach AG, Longbottom DA, Nesi M, Scott JS, Storer RI, Taylor SJ (2000) Multi-step organic synthesis using solid-supported reagents and scavengers: a new paradigm in chemical library generation. J Chem SocP erkin Trans 1 3815–4195

Ley SV, Takemoto T, Yasuda K (2001) Solid-supported reagents for the oxidation of aldehydes to carboxylic acids. Synlett 1555–1556

Ley SV, Lumeras L, Nesi M, Baxendale IR (2002a) Synthesis of trifluoromethyl ketones using polymer-supported reagents. Comb Chem High Throughput Screening 5:195–199

Ley SV, Baxendale IR, Brusotti G, Caldarelli M, Massi A (2002b) Solid-supported reagents for multi-step organic synthesis: preparation and application. Farmaco 57:321–330

Ley SV, Ramarao C, Gordon RS, Holmes AB, Morrison AJ, McConvey IF, Shirley IM, Smith SC, Smith MD (2002c) Polyurea-encapsulated palladium(II) acetate: a robust and recyclable catalyst for use in conventional and supercritical media. J Chem Soc Chem Commun 1134–1136

Ley SV, Baxendale IR, Myers RM (2005) Polymer-supported regents scavengers and catch-and-release techniques for chemical synthesis. In: Taylor J, Triggle D (eds) Comprehensive medicinal chemistry II, Vol. 3, Drug Discovery Technologies. Elsevier, Oxford, pp 79–836

Ley SV, Baxendale IR, Myers RM (2006) The use of polymer supported reagents and scavengers in the synthesis of natural products. In: Boldi AM (ed) Combinatorial synthesis of natural product-based libraries. CRC Press Boca-Raton, pp 131–163

Luckarift HR, Nadeau LJ, Spain JC (2005) Continuous synthesis of aminophenols from nitroaromatic compounds by combination of metal and biocatalyst. J Chem Soc Chem Commun 383–384

Miura M (2004) Rational ligand design in constructing efficient catalyst systems for suzuki-miyaura coupling. Angew Chem Int Ed 43:2201–2203

Nüchter M, Ondruschka B, Bonrath W, Gum A (2004) Microwave-assisted synthesis – a critical technology overview. Green Chem 6:128–141

Parlow JJ, Devraj RV, South MS (1999) Solution-phase chemical library synthesis using polymer-assisted purification techniques. Curr Opin Chem Biol 3:320–336

Pershichini PJ (2003) Carbon-Carbon Bond Formation via Boron Mediated Transfer. Curr Org Chem 7:1725–1736

Ramarao C, Ley SV, Smith SC, Shirley IM, DeAlmeida NJ (2002) Encapsulation of palladium in polyurea microcapsules. Chem Soc Chem Commun 1132–1133

Saaby S, Baxendale IR, Ley SV (2005) Non-metal-catalysed intramolecular alkyne cyclotrimerization reactions promoted by focused microwave heating in batch and flow modes. Org Biomol Chem 3:3365–3368

Sands M, Haswell SJ, Kelly SM, Skelton V, Morgan DO, Styring P, Warrington B (2001) The investigation of an equilibrium dependent reaction for the formation of enamines in a microchemical system. Lab Chip 1:64–65

Schwartz MA, Holton RA (1970) Intramolecular oxidative phenol coupling. II. Biogenetic-type synthesis of (+-)-maritidine. J Am Chem Soc 92:1090–1092

Sherrington DC (2001) Polymer-supported reagents, catalysts, and sorbents: evolution and exploitation – a personalized view. J Polym Sci Pol Chem 39:2364–2377

Shuttleworth SJ, Allin SM, Sharma PK (1997) Functionalised polymers: recent developments and new applications in synthetic organic chemistry. Synthesis 1217–1239

Shuttleworth SJ, Allin SM, Wilson RD, Nasturica D (2000) Functionalised polymers in organic chemistry: part 2. Synthesis 8:1035–1074

Siu J, Baxendale IR, Ley SV (2004) Microwave assisted Leimgruber–Batcho reaction for the preparation of indoles, azaindoles and pyrroylquinolines. Org Biomol Chem 160

Storer RI, Takemoto T, Jackson PS, Ley SV (2003) A total synthesis of epothilones using solid-supported reagents and scavengers. Angew Chem Int Ed 42:2521–2525

Storer RI, Takemoto T, Jackson PS, Brown DS, Baxendale IR, Ley SV (2004) Multi-step application of immobilized reagents and scavengers: a total synthesis of epothilone C. Chem Eur J 10:2529–2547

Sussman S (1946) Ind Eng Chem 38:1228

Thomas JR, Faucher FJ (2000) Thermal modeling of microwave heated packed and fluidized bed catalytic reactors. Microw Power Electromagn Energy 35:165–174

Thompson LA (2000) Recent applications of polymer-supported reagents and scavengers in combinatorial, parallel, or multistep synthesis. Curr Opio Chem 4:324–337

Toteja RSD, Jangida BL, Sundaresan M, Venkataramani B (1997) Water sorption isotherms and cation hydration in dowex 50w and amberlyst-15 ion exchange resins. Langmuir 13:2980–2982

Vickerstaff E, Warrington BH, Ladlow M, Ley SV (2004) Fully automated polymer-assisted synthesis of 1,5-biaryl pyrazoles. J Comb Chem 6:332–339

Wathey B, Tierney JP, Lidstrom P, Westman J (2001) Microwave-assisted organic synthesis—a review. Tetrahedron 57:9225–9283

Wathey B, Tierney JP, Lidstrom P, Westman J (2002) The impact of microwave-assisted organic chemistry on drug discovery. Drug Discovery Today 7:373–380

Watts P, Haswell SJ (2003a) Microfluidic combinatorial chemistry. Curr Opin Chem Biol 7:380–387

Watts P, Haswell SJ (2003b) Continuous flow reactors for drug discovery. Drug Discov Today 8:586–593

Wison NS, Sarko CR, Roth GP (2004) Development and applications of a practical continuous flow microwave Cell. Org Proc Res Dev 8:535–538

Yu J-Q, Wu H-C, Ramarao C, Spencer JB, Ley SV 2003) Transfer hydrogenation using recyclable polyurea-encapsulated palladium: efficient and chemoselective reduction of aryl ketones. J Chem Soc Chem Commun 678–679

Zu ZP, Chuang KT (1996) Kinetics of acetic acid esterification over ion exchange catalysts. Can J Chem Eng 74:493–500

Ernst Schering Foundation Symposium Proceedings, Vol. 3, pp. 187–203
DOI 10.1007/2789_2007_034

Published Online: 23 May 2007

Solid-Phase Supported Synthesis: A Possibility for Rapid Scale-Up of Chemical Reactions

M. Meisenbach(✉), T. Allmendinger, C.-P. Mak

Chemical and Analytical Development, Novartis Pharma AG, 4002 Basel, Switzerland
email: *mark.meisenbach@novartis.com*

Abstract. The direct scale-up of a solid-phase synthesis has been demonstrated with 4-(2-amino-6-phenylpyrimidin-4-yl)benzamide and an arylsulfonamido-substituted hydroxamic acid derivative as examples. These compounds were obtained through combinatorial chemistry and solution-phase synthesis was used in parallel to provide a comparison. By applying highly loaded polystyrene-derived resins as the solid support, a good ratio between the product and the starting resin is achieved. We have demonstrated that the synthesis can be scaled up directly on the solid support, successfully providing the desired compounds easily and quickly in sufficient quantities for early development demands.

1 Introduction

Solid-phase synthesis for small organic compounds is widely used in research and academia for rapid lead finding and lead optimisation. The number of hits obtained from combinatorial methods is expected to increase in the future and consequently numerous compounds as potential drug candidates will enter into development. Process chemists are then faced with the problem of rapidly delivering the first amounts necessary for early preclinical and clinical studies. If the compounds were obtained by a solid-phase approach, solution-phase procedures for their synthesis might not be readily available. To tackle this problem, we were interested in gaining experience in the direct scale-up of solid-phase syntheses without investing resources searching for a solution-phase alternative. An added advantage in this approach is inherent: the potential time savings since multi-step procedures could be conducted without isolation and purification of intermediates. Our group had previously reported on this topic (Meisenbach et al. 2003) and very few examples are found in the literature (Raillard et al. 1999; Prühs et al. 2006). Herein we will expand upon our earlier work with additional examples.

With the following examples, we will investigate and discuss the following: (a) Scale-up phenomena of different solid phase reactions and the corresponding on-bead analytics; (b) the effect of loading (an equivalent to concentration in solution-phase chemistry); (c) comparison with the solution-phase alternative and (d) the synthesis and use of new trityl linkers.

2 Results and Discussion

The first example describes the synthesis of a pyrimidine derivative. Starting from α,β-unsaturated ketones (see Schemes 1, 8), a library of different heterocycles was prepared in research (Felder and Marzinzik 1998). In preparation for any large-scale synthesis, the availability of starting materials is always considered (Lee and Robinson 1995). For this work, we had to replace Rink amide resin **B** (Rink 1987), which was used by our colleagues in research for the synthesis of pyrimidine **1** due to its unavailability in large quantities (see Fig. 1). It was replaced with the Rink amide acetamido resin **4**, which is well established in peptide amide synthesis (Bernatowicz et al. 1989) and easily accessible.

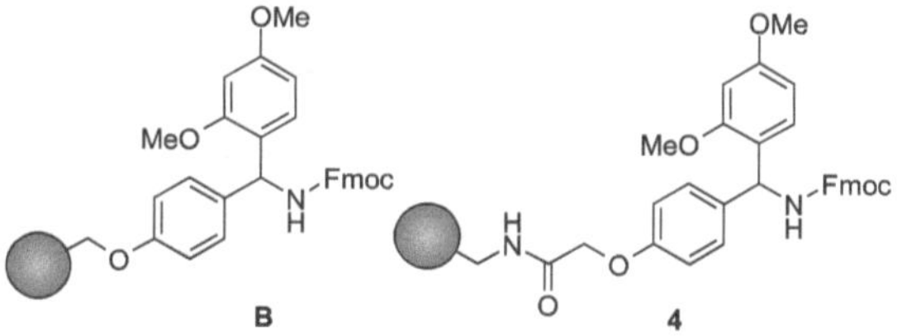

Fig. 1. Structural comparison of the Rink-linker compared to the linker used for the scale-up

Thus (see Scheme 1), 1.5 equiv. of the acetic acid derivative **3** was coupled to aminomethyl polystyrene (AMPS **2a**, 1.54 mmol/g) by standard amide-forming conditions. After complete acylation as indicated by a negative Kaiser-test (Kaiser et al 1970) for unreacted amino functions, resin **4a** was checked for its loading using well-elaborated UV-techniques (Meienhofer et al. 1979). After cleaving the Fmoc group with 20% diethylamine in DMF, 4-carboxybenzaldehyde **5** was coupled (diisopropyl carbodiimide DIC, HOBt). Increasing the reaction time allowed the reduction of the amount of building block **5** by a factor of 10 as compared to the original protocol. The structure of the resulting solid-phase-supported aldehyde **6a** was confirmed by IR spectroscopy (Yan et al. 1999), which showed a strong C=O band at 1,710 cm^{-1} and the characteristic C–H band at 2,730 cm^{-1}.

The subsequent Claisen-Schmidt reaction was originally performed on a 10-μmol scale using 20-fold excess of both acetophenone and LiOH to achieve complete formation of the chalcone **8**. This result could be verified on a small scale; however, employing the same conditions on a 35-mmol scale resulted in no conversion even after 22 h, as revealed by IR spectroscopy. By cleaving a resin sample with 20% TFA in dichloromethane, only *p*-formylbenzamide **11** was detected by HPLC. This result may be explained by the low solubility of LiOH in DME under dry/aprotic conditions. Therefore, a small amount of EtOH was added, which initiated a fast reaction (Chiu et al. 1999) and the formation of the desired chalcone **8** together with 20% of the Michael adduct **10** (Fig. 2). This was confirmed by sample cleavage from the resin and LC-MS analysis. Short reaction screening resulted in considerable im-

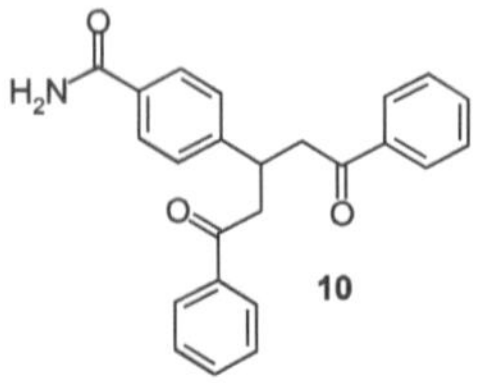

Fig. 2. Structure of the Michael-adduct

provements: using less acetophenone **7** and LiOH (1.5 and 0.5 equivalents, respectively) and a mixed solvent system of THF and methanol gave complete conversion within only 1 h, forming only 5% of Michael adduct as by-product. To get more insight into what exactly was going on, online Raman spectroscopy was used to monitor the conversion of this Claisen-Schmidt reaction (Dyson et al. 2000) (Fig. 3). The measurements indicate that the optimum reaction time is 50 min before the signal at 1,608 cm^{-1} starts to decrease, which is when the product starts to be consumed because of the Michael addition (see Fig. 3).

The final conditions were scaled up with no problems: chalcone **8a** and guanidine (liberated from its hydrochloride with sodium ethoxide) were heated in dimethyl acetamide while bubbling air through the mixture to form pyrimidine **9a**. After complete conversion (16 h), the product was cleaved from the support (20% trifluoroacetic acid in DCM). Pure 4-(2-amino-6-phenyl-pyrimidin-4-yl)benzamide **1** was obtained as its trifluoroacetate salt upon evaporation of the filtrate and recrystallisation of the residue from ethanol/water in 56% overall yield based on the solid phase attached 4-carboxybenzaldehyde **6a**.

After isolating the product of a solid phase synthesis, the support (resin + linker) is usually discarded as waste, although successful examples of its reuse in further synthetic cycles are known with trityl type linkers (Frechet and Haque 1975). To reduce both volume of operation and amount of waste, the loading of the resin (quantified as millimoles of functionality per gram) has to be increased. Besides theoretical limitations (for polystyrene this is reached when every phenyl ring is substituted by the linker), there may be practical boundaries for using highly loaded resins in solid-phase supported synthesis. This issue was studied

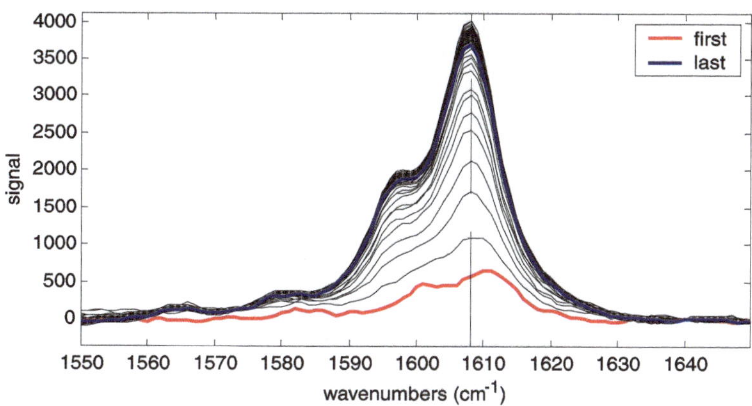

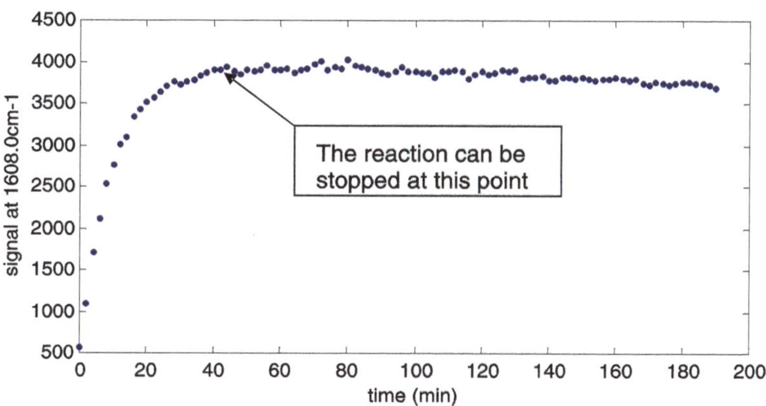

Fig. 3. a Raman spectra (preprocessed); this peak predominantly shows formation of product. **b** Spectral changes over time shown for the wave number of maximum signal (1608 cm^{-1})

in detail with the preparation of **1**, where the resin loading was varied. This was also another reason to switch from the ether-bonded Rink-linker **B** to the acetic acid derivative **4**, which can be assembled from **3** and differently loaded aminomethylated polystyrene resins (AMPS, **2**). AMPS is commercially available with loading up to 2.9 mmol/g **2b** (NovaBiochem, Switzerland). By adapting and optimising the procedure developed by Adams (Adams et al. 1998), higher loaded resins were prepared. Thus polystyrene cross-linked with 1% divinylbenzene was reacted with *N*-chloromethyl phthalimide **12** in dichloromethane using iron (III) chloride as Friedel-Crafts catalyst. If the reaction is performed at 25°C a colourless product **13** is obtained (Scheme 3), in con-

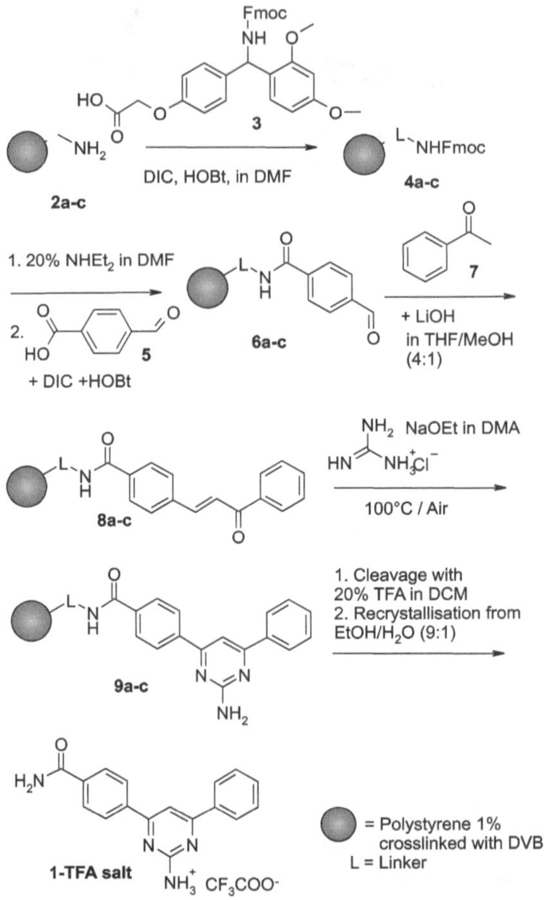

Scheme 1. SPSS of the pyrimidine derivative

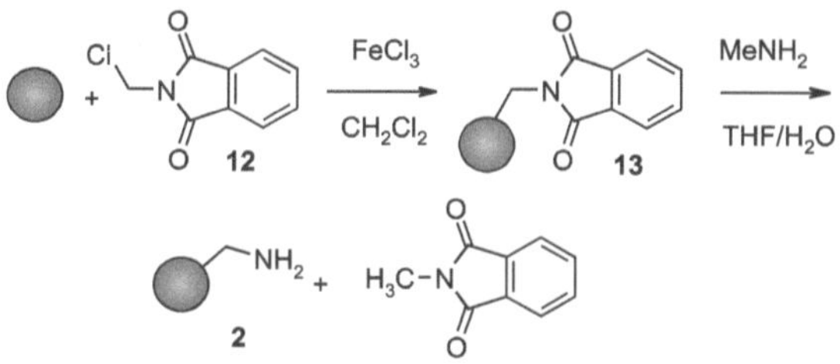

Scheme 2. Synthesis scheme of the AMPS

trast to dark yellow resins obtained when working under reflux. The phthaloyl protective group was removed by means of aminolysis using methylamine in water/THF in order to avoid the common but toxic hydrazine and dioxane. The desired loading can be obtained by adjusting the ratio of PS, chloromethyl phthalimide and $FeCl_3$. We synthesised AMPS with up to 4.48 mmol/g (**2c**) corresponding to two out of three phenyl rings being aminomethylated. We have not extended the investigation to the highest possible loading of AMPS. A resin with up to 7.3 mmol/g (96% of phenyl-rings on polystyrene are substituted) has been described previously (Zikos and Ferderigos 1995), but its application has not been reported so far.

To study the effect of loading, the synthesis depicted in Scheme 1 was repeated using different resin loadings. The conditions during reactions on higher loaded resins (**2b** and **2c**) as compared to the initially used **2a** did not need any adjustments. The quality of the product was comparable and the ratio of product to solid support was increased (Table 1).

Without the experience and the equipment to perform solid-phase supported syntheses on a larger scale, chemists explore alternative routes utilising solution-phase chemistry. To compare the direct scale-up on solid support and to evaluate the advantages and disadvantages of both approaches, pyrimidine **1** was prepared in solution phase as well. For this relatively small and simple molecule, similar chemistry was applied (see Scheme 3) with some interesting results arising.

Heating 4-carboxybenzaldehyde **5** in excess thionyl chloride, which is described to form acid chloride **14** (Graffner-Nordberg et al. 1998),

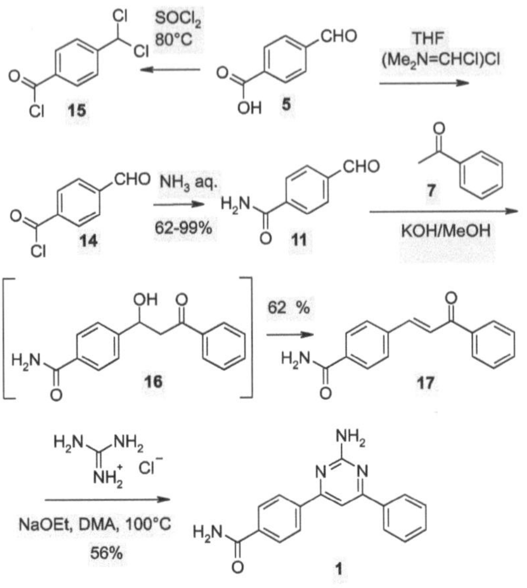

Scheme 3. Synthesis of the pyrimidine derivative in liquid phase

Table 1 Solid phase synthesis of pyrimidine **1-TFA salt** starting with resins of different loading

Loading of AMPS, scale of synthesis	Mass of **9** (g)	Yield of crude **TFA salt**		Purity (area %)*	Overall yield and purity after recrystallisation of **1-TFA salt**	
		in (g)	in (%)		in (%)	in (area %)*
2a, 1.54 mmol/g 40 mmol, 26 g	49	14.9	92	84	56	>98
2b, 2.86 mmol/g 70 mmol, 24.5 g	61.39	17.8	66	80	47	>97
2c, 4.48 mmol/g 100 mmol, 22.3 g	74.9	31.8	82	86	55	>98

* RP-HPLC

Table 2 Comparison of the liquid and solid phase synthesis of the pyrimidine derivative 1

	Yield	# of steps	Time of manufacturing
Liquid phase	39%	4	3d
Solid phase	73%	4 + cleavage + crystallisation	3d

actually afforded the trichloro-compound **15** instead. Therefore the milder Vilsmeier reagent was used, followed by aminolysis in order to obtain **11**. Due to its solubility in water, variable yields of **11** were obtained depending on the workup procedure employed. The Aldol condensation with acetophenone under standard conditions gave chalcone **17** containing 10% of the intermediate **16**. Its structure was revealed by independent synthesis (Ohki et al. 1988), to compare the NMR-data and chromatographic properties. Hydroxyketone **16** turned out to be hardly soluble (even in dimethyl sulfoxide), and precipitated with **17**. These phenomena indicate the advantage of solid-phase supported synthesis when intermediates are involved, possessing either very low solubility in organic solvents (like **16**) or high aqueous solubility (like **11**). Finally the condensation with guanidine and the following oxidation afforded pyrimidine **1** in moderate yield. The results are compared in Table 2.

As second example for the scale-up of solid-phase reactions directly on solid support, we chose an arylsulfonamido-substituted hydroxamic acid derivative stemming from the matrix metalloproteinase inhibitor library (MMP) of our research colleagues (Breitenstein et al. 2001). In this case, there was already a solution-phase synthesis available for comparison (see Scheme 4). The synthesis starts with the imine formation of a benzaldehyde **18** with the glycine methyl ester, which is then reduced to the benzylamine **20** using sodium borohydride in methanol/THF (2:1). The sulfonamide formation is carried out in dioxane/H_20 (2:1) with triethylamine as the base and after neutralisation and evaporation the product **21** can be crystallised from tert. butylmethyl ether. After deprotection with LiOH, the acid is activated by treatment with oxalyl chloride and finally converted into the hyroxamic acid **23** in 33.7% yield overall.

The SPSS following this route is illustrated in Scheme 5. In the first step, Fmoc-protected glycine is attached to a trityl-derived hydroxyl

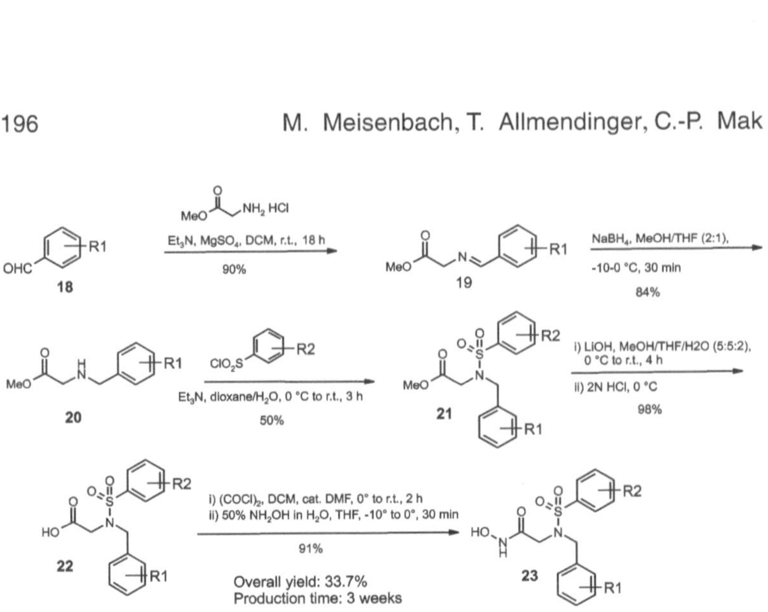

Scheme 4. Synthesis of a MMP inhibitor, a hydroxamic acid derivative in liquid phase

amine resin using DIC and HOBt. A sterically demanding linker was selected because of the reactivity of the hydroxamic acid's nitrogen. After the cleavage of the Fmoc group, the polymer bounded benzylamine **26** is synthesised via reductive amination (see also Table 3). The sulfonamide formation is performed in DCM using *N*-methylmorpholine as base, followed by the final cleavage of the product from the resins using TFA in different concentrations in DCM depending on the linker used.

At first glance, the synthesis of **23** seems to be straightforward. However, some difficulties were encountered, which are described below.

a. The monitoring of the imine formation c and reduction d by simple cleavage and HPLC-analysis is impossible due to instability of the intermediates under acidic conditions. IR-monitoring (C=X-bands) was ineffective because a CO-N-functionality is always present (Table 3, no. 1).
b. **23** Itself turned out to be labile towards the harsher cleavage conditions during sample evaporation (Table 3, no. 2a). Only a small scale run with linker **24a** was successful when the solution after cleavage was immediately neutralized; otherwise debenzylation occurred forming **28** (Table 3, no. 2b).

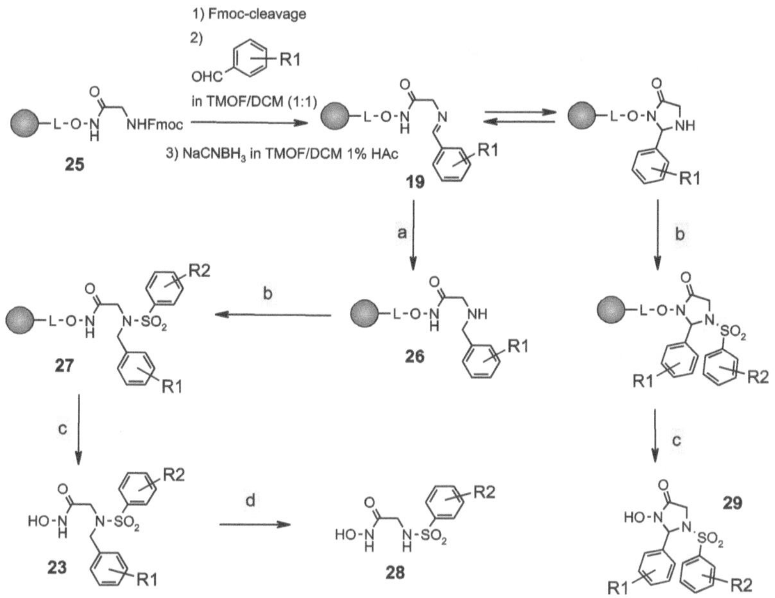

Scheme 5. SPS synthesis of the hydroxamic acid derivative via reductive amination **a** Reduction, **b** Cl-SO_2-Ph-R2, NMM, DCM 3–4 h **c** 10% or 1–2% TFA in DCM with direct neutralisation with sat. $NaHCO_3$, **d** without neutralisation

c. Analogous to the solution-phase results (Harmon et al. 1970), imines from glycine hydroxamic acids (like **19**) may form imidazolinones; these would not be reduced but sulfonylated in the following step forming **29** (Scheme 6). Indeed, the outcome of the whole sequence turned out to be highly dependent on the conditions of the reduction step a. We tried to substitute the highly toxic sodium cyanoborohydride by $NaBH(OAc)_3$ or $NaBH_4$, but this was unsuccessful (Table 3, nos. 4, 5). Only when using $NaCNBH_3$ in a mixture of dichloromethane, trimethyl orthoformate containing 1% acetic acid was the desired **23** obtained. Without acetic acid (which we wanted to avoid in combination with $NaCNBH_3$) or using different solvents (DMF or THF), the reduction was not successful and **29** was isolated (Table 3, nos. 3, 6).

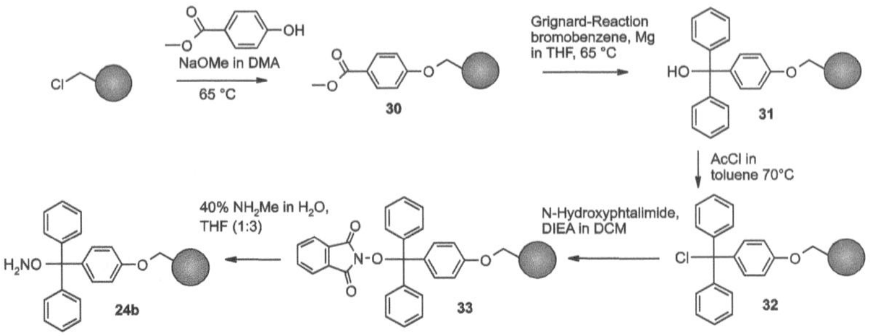

Scheme 6. Synthesis scheme for the preparation of the 4-benzyloxy-trityl hydroxylamine

For these reasons, an alternative route and more acid labile linkers compared to p-carboxy trityl linker **24a** initially used were sought, to avoid high concentrations of TFA for the final cleavage. The synthesis of the alkoxysubstitued linkers **24b** (Meisenbach and Voelter 1997) and **24c**, which can be synthesised directly on the solid support in five steps, offer the possibility of linkers with tailor-made stability.

In the case of the benzyloxytrityl-hydroxylamine resin **24b**, the starting material is the Merrifield resin, chloromethylated polystyrene (1.7 mmol Cl $\times$ g^{-1}) which is treated with 3 equiv. of 4-hydroxy-methyl benzoate in DMA with sodium methylate as base. The Beilstein test is used for monitoring and microanalyses showed the absence of chloride (Scheme 6).

Resin **24c** is a far cheaper alternative also allowing a facile variation of its loading (Scheme 7). Starting with inexpensive 1% cross-linked polystyrene, a versatile linker was prepared following the protocol of Fréchet (Fréchet and Nuyens 1976). The Friedel-Crafts benzoylation is carried out using DCM instead of highly toxic and highly inflammable carbon disulfide at 5°C with 0.66 equiv. $AlCl_3$ and 0.6 equiv. benzoylchloride affording a resin where two-thirds of the phenyl rings are functionalised. The trityl system is then generated by a Grignard reaction with 10 equiv. of the corresponding bromo compounds and 10 equiv. magnesium in THF refluxing for 24 h. The structures were confirmed by the disappearance of the CO band in the IR spectrum at 1,710 cm^{-1} and the appearance of a strong OH band at 3,500 cm^{-1}, indi-

Table 3 Solid phase synthesis of **23** *via* reductive amination

No. of Exp.	Reaction	Linker	Red. agent	Solvent	Cleavage conditions	Comp. **23** HPLC [A%]
1	Red. alkylation	**24a**	$NaCNBH_3$	TMOF/DCM 1% HAc	20% TFA/DCM	2%
2a	1. Imine formation 2. Reduction	**24a**	$NaCNBH_3$	1. TMOF/DCM 2. TMOF/DCM, 1% HOAc	20% TFA/DCM	48% 28% **28**
2b					10% TFA/DCM with direct neutralisation into $NaHCO_3$ Sol.	80%
3	1. Imine formation 2. Reduction	**24b**	$NaCNBH_3$	1. TMOF/DCM 2. TMOF/DCM	1% TFA/DCM with direct neutral. into $NaHCO_3$ solution	79% **29**
4	1. Imine formation 2. Reduction	**24a**	$NaBH(OAc)_3$	1. TMOF/DCM 2.1% HOAc in DMF	10% TFA/DCM with direct neutralisation into $NaHCO_3$ solution	40%
5	1. Imine formation 2. Reduction	**24a**	$NaBH_3$	1. TMOF/DCM 2. THF/MeOH (3:1)	10% TFA/DCM with direct neutralisation into $NaHCO_3$ solution	54%
6	1. Imine formation 2. Reduction	**24b**	$NaCNBH_3$	1. TMOF/DCM 2. 1% HOAc in DMF	1% TFA/DCM with direct neutralisation into $NaHCO_3$ solution	80% **29**

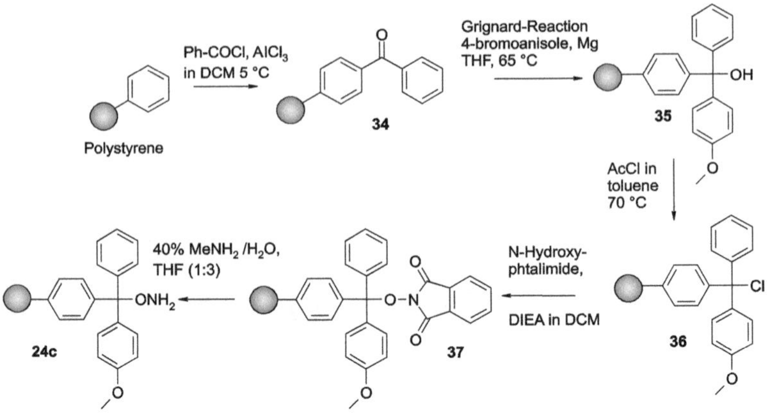
Ph-COCl, AlCl3
in DCM 5 °C
Polystyrene
34
Grignard-Reaction
4-bromoanisole, Mg
THF, 65 °C
OH
35
AcCl in
toluene
70 °C
O
Cl
36
N-Hydroxy-
phtalimide,
DIEA in DCM
O-N
37
40% MeNH2 /H2O,
THF (1:3)
ONH2
24c

Scheme 7. Synthesis scheme for the preparatin of the 4-methoxy-trityl hydroxylamine linker

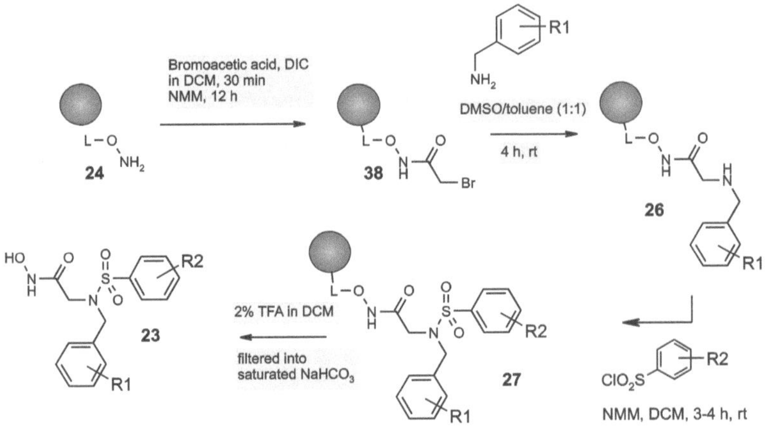
L-O
NH2
24
Bromoacetic acid, DIC
in DCM, 30 min
NMM, 12 h
38
Br
R1
NH2
DMSO/toluene (1:1)
4 h, rt
26
ClO2S
R2
NMM, DCM, 3-4 h, rt
27
2% TFA in DCM
filtered into
saturated NaHCO3
HO
23

Scheme 8. New route for the SPSS of hydroxamic acid derivative by nucleophilic substitution

cating that the reaction had gone to completion. The OH-functionalized resins were transformed into the orange halide when reacted with AcCl in toluene at 70°C for 1 h. Finally, upon treatment with 3 eq. of N-hydroxyphthalimide and Hünig's base in dichlormethane, the chloro-

Table 4 Solid phase synthesis of **23**

PS	**24c** resin	Scale of synthesis and mass of starting resin **24c**	Scale of synthesis and mass of starting resin **38**	Crude **23** after cleavage with 2%TFA/DCM	**23** after crystallisation from EtOAc/heptane
				Yield based on **38**, purity HPLC	
292 g	556 g,	201.5 g/ 345 mmol	235 g/ 0.93 mmol/g, 220 mmol	75.26 g, 77.2%, 86.3A%	42.4 g, 43.5%, 99A%
	1.71 mmol/g	268 g/ 458 mmol	315 g/ 0.74 mmol/g, 233 mmol	83.08 g, 80.5%, 87.1A%	54.5 g, 52%, 99.1A%

resins are converted to **33** and **37**, respectively. The cleavage of phthaloyl protective group was carried out as described above for the AMPS resins and can be easily monitored by the disappearance of the CO band in the IR spectrum at 1,710 cm^{-1}.

With these solid supports in hand, we turned our attention to a new route to the synthesis of our target molecule **23** (Scheme 8). The tricky reductive amination should be replaced by an N-alkylation. To that end, bromoacetic acid is attached to **24c** using DIC and Hünig's base followed by the nucleophilic substitution with the corresponding benzylamine in DMSO/toluene (1:1), which can be easily monitored by the Beilstein test, followed by sulfonamide formation in DCM using *N*-methylmorpholine as base. For the final cleavage, 2% TFA in DCM is used and the resulting solution is filtered in a saturated $NaHCO_3$ solution to neutralise the acid before evaporation of the solvent. The crude product was then crystallised from ethyl acetate/heptane to yield the desired product in 27% yield overall and 99A% HPLC purity (see Table 4).

3 Conclusion

In summary, we demonstrated the possibility of the fast scale-up of solid-phase supported synthesis. In one case (pyrimidine derivate), the research protocols could be used directly with only minor modifications. In another case, an alternative synthesis was needed for a successful and efficient scale-up. For these examples with only a few reaction steps, the production time on solid support is comparable with a synthesis in solution but the overall yield (47%–56%) exceeds that obtained by solution phase synthesis (21%–34%). A further achievement is the accessibility of highly loaded AMPS and trityl resins to increase the product/resin ratio and the volume productivity. Combining these factors, the competitive position of a fast up-scale of given SPSS-protocols compared to the time-consuming development of a new solution phase synthesis for small compounds is considerably improved, in particular when the time savings is considered, one of the most important factors in the early stage of drug development.

References

Adams JH, Cook RM, Hudson D, Jammalamadaka V, Lyttle MH, Songster MF (1998) A reinvestigation of the preparation, properties, and applications of aminomethyl and 4-methylbenzhydrylamine polystyrene resins. J Org Chem 63:3706–3716

Bernatowicz MS, Daniels SB, Köster H (1989) A comparison of acid-labile linkage agents for the synthesis of peptide C-terminal amides. Tetrahedron Lett 30:4645–4648

Breitenstein W, Hayakawa K, Iwasaki G, Kanazawa T, Kasaoka T, Koizumi S, Matsunaga S, Nakajima M, Sakaki J (2001) Preparation of arylsulfonamido-substituted hydroxamic acid derivatives as MMP2 inhibitors. Int Pat Appl WO 2001010827

Chiu C, Tang Z, Ellingboe JW (1999) Solid-phase synthesis of 2,4,6-trisubstituted pyridines. J Comb Chem 1:73–77

Dyson RM, Helg AG, Meisenbach M, Allmendinger T (2000) In situ observation of the progress of a large-scale on-bead reaction. 11th International Symposium on Pharmaceutical and Biomedical Analysis, Basle, Switzerland, p 007

Felder ER, Marzinzik AL (1998) Key intermediates in combinatorial chemistry: access to various heterocycles from α,β-unsaturated ketones on solid phase. J Org Chem 63:723–727

Fréchet JMJ, Haque KE (1975) Use of polymers as protecting groups in organic synthesis. II. Protection of primary alcohol functional groups. Tetrahedron Lett 16:3055–3056

Fréchet JMJ, Nuyens LJ (1976) Use of polymers as protecting groups in organic synthesis. III. Selective functionalisation of polyhydroxy alcohols. Can J Chem 54:926–934

Graffner-Nordberg M, Sjödin K, Tunek A, Hallberg A (1998) Synthesis and enzymatic hydrolysis of esters, constituting simple models of soft drugs. Chem Pharm Bull 46:591–601

Harmon RE, Rizzo VL, Gupta SK (1970) Synthesis of 3-hydroxy-4-imidazolidinones. J Heterocyc Chem 7:439–442

Kaiser E, Colescott RL, Bossinger CD, Cook PI (1970) Color test for detection of free terminal amino groups in the solid-phase synthesis of peptides. Anal Biochem 34:595–598

Lee S, Robinson G (1995) Process development: fine chemicals from grams to kilograms. Oxford University Press, Oxford

Meienhofer J, Waki M, Heimer EP, Lambros TJ, Makofske RC, Chang CD (1979) Solid phase synthesis without repetitive acidolysis. Preparation of leucyl-alanyl-glycyl-valine using 9-fluorenylmethyloxycarbonylamino acids. Int J Pep Prot Res 13:35–42

Meisenbach M, Voelter W (1997) New methoxy-substituted tritylamine linkers for the solid phase synthesis of protected peptide amides. Chem Lett 1265–1266

Meisenbach M, Allmendinger T, Mak CP (2003) Scale-up of the synthesis of a pyrimidine derivative directly on solid support. Org Proc Res Dev 7:553–558

Ohki H, Wada M, Akiba K (1988) Bismuth trichloride as a new efficient catalyst in the aldol reaction. Tetrahedron Lett 29:4719–4720

Prühs S, Dinter C, Blume T, Schütz A, Harre M, Neh H (2006) Upscaling the solid-phase synthesis of a tetrahydrocarbazole in chemical development. Org Proc Res Dev 10:441–445

Raillard SP, Ji G, Mann AD, Baer TA (1999) Fast scale-up using solid-phase chemistry. Org Proc Res Dev 3:177–183

Rink H (1987) Solid phase synthesis of protected peptide fragments using a trialkoxydiphenyl-methylester resin. Tetrahedron Lett 28:3787–3790

Yan B, Gremlich HU, Moss S, Coppola GM, Sun Q, Liu LJ (1999) Comparison of various FTIR and FT Raman methods: applications in the reaction optimisation stage of combinatorial chemistry. J Comb Chem 1:46–54

Zikos CC, Ferderigos NG (1995) Preparation of high capacity aminomethyl-polystyrene resin. Tetrahedron Lett 36:3741–3744

Ernst Schering Foundation Symposium Proceedings, Vol. 3, pp. 205–240
DOI 10.1007/2789_2007_035

Published Online: 24 May 2007

Microstructured Reactors for Development and Production in Pharmaceutical and Fine Chemistry

V. Hessel(✉), P. Löb, U. Krtschil, H. Löwe

Department of Chemical Engineering and Chemistry,
Eindhoven University of Technology, STW 1.35, PO Box 513, 5600 MB Eindhoven, The Netherlands
email: *v.hessel@tue.nl*

Abstract. The true potential of microprocess technology for process intensification is not yet fully clear and needs to be actively explored, although more and more industrial case stories provide information. This paper uses a shortcut cost analysis to show the major cost portions for processes conducted by microstructured reactors. This leads to predicting novel chemical protocol conditions, which are tailored for microprocess technology and which are expected to highly intensify chemical processes. Some generic rules to approach this are termed new process windows, because they constitute a new approach to enabling chemistry. Using such process intensification together with scaled-out microstructured reactors, which is demonstrated by the example of gas–liquid microprocessing, paves the road to viable industrial microflow processes. Several such commercially oriented case studies are given. Without the use of new process windows conditions, microprocess technology will probably stick to niche applications.

1 Microreactors

Microreactors are reactors made by microfabrication that typically contain microchannels as flow-through elements for continuous processing (Hessel et al. 2004a, 2005; Ehrfeld et al. 2000; Jensen 1999, 2001; Fletcher et al. 2002; Haswell and Watts 2003; Gavriilidis et al. 2002; Jähnisch et al. 2004; Kolb and Hessel 2004; Pennemann et al. 2004). Microreactors in their most compact version comprise chemical chip systems, i.e., made by silicon micromachining, glass photoetching, and plastic engineering. Fist- and shoebox-sized laboratory and pilot units, respectively, are adequately termed microstructured reactors and conveniently made by means of steel etching, microelectro-discharge machining, and precision engineering such as milling, turning, or punching. In a few cases, much larger units were developed with outer dimensions in the meter range, in one case extended to even storey size (Markowz et al. 2005).

Today, specialized microstructured equipment is available for many chemical engineering operations such as gas, liquid, gas–liquid, and liquid–liquid processing, largely without involving solids (Hessel et al. 2004a, 2005d; Ehrfeld et al. 2000; Jensen 1999, 2001; Fletcher et al. 2002; Haswell and Watts 2003; Gavriilidis et al. 2002; Jähnisch et al. 2004; Kolb and Hessel 2004; Pennemann et al. 2004). This is further expanded to catalytic operations by insertion of a catalyst on carriers, most often as coated layers or mini fixed beds, or as thin pure catalyst films (Kiwi-Minsker and Renken 2005; Kolb and Hessel 2004).

For many single-phase liquid operations, which comprise the majority of organic reactions, the connection of only two fluidic elements is sufficient to replace the commonly used mini-batch reactors, i.e., the typical stirred glass-lined, multiple-necked vessels with reflux condensers (Hessel et al. 2004a; Fletcher et al. 2002; Haswell and Watts 2003; Jähnisch et al. 2004; Pennemann et al. 2004). A micromixer is followed by an integrated microreactor–micro-heat exchanger or a more simply heated microreactor, providing the needed mixing, heating/cooling, and reaction time functions. At the edge of very fast reactions, the micromixer also operates as a reactor, and the reaction is completed here, with no need for any other unit. At the edge of slower reactions, typically in the range of several minutes to as much as an hour, the micromixer is connected to a long residence unit, typically a capillary, and often only the capillary itself is used, with no need for a micromixer. Simple mixing tees then bring the reactants together, or even premixed solutions are used.

Different types of micromixers exist based on different passive mixing principles (Hessel et al. 2004b). One class utilizes only diffusion at the micro scale by creating the finest geometric fluidic elements (jets, lamellae, etc.). A second class generates a recirculation flow to use convection, and then diffusion. A few micromixers are based on turbulence, since channel-free jets collide frontally at very high velocity, or flows at sufficiently high Reynolds numbers are achieved in comparatively large mixing channels. Besides micromixers, compact fluidic or electrically driven micro-heat exchangers were developed (Schubert et al. 2001). In the field of gas–liquid systems, falling film microreactors, Taylor-flow and annular-flow channel contactors, and mini-trickle bed reactors are most often used (Hessel et al. 2005a). For liquid–liquid operation, slug-flow capillary reactors are one suitable reactor concept (Burns and Ramshaw 2001). Special dual- and triple-feed channel architectures can create droplets (Link et al. 2004; Utada et al. 2005; Joanicot and Ajdari 2005) or particles (Xu et al. 2005) of unique or uniform shape or with uniformly structured morphology or chemical composition, which are often changed by simple process parametric variation. This stems from the well-defined flow patterns and concentration and temperature fields of the microflows.

From the very beginning, continuous reactor concepts, an alternative to the truly microfabricated reactors, were used, for example, static meso-scaled mixers or HPLCs and other smart tubing (see Iwasaki et al. 2006 for an example). This completed functionality by filling niches not yet covered by microfabricated reactors or even by replacing the latter as a more robust, more easily accessed or more inexpensive processing tool. Further innovative equipment, coming from related developments in the process intensification field, is another source: e.g., structured packings such as fleeces, foams, or monoliths.

2 Microreactor Processes and Plants

In the last few years, the corresponding process technology and plants for microstructured reactors have been established, both by new customized process-flow environments from specialized providers (see Fig. 1) and by retrofit into existing plants at the company site.

This named the disciplines of microprocess technology and engineering. The microprocess technology plants can have a shorter time-to-market, better scale-out predictability, a higher degree of modularity, less stringent legislation, and are principally transportable or easy to dismount (Deibel 2006). Thus, even industry captains and leaders draw the conclusion that, sooner or later, parts of the existing plants will be replaced by the new technology and that the classical world-scale plant tends to be a phase-out model (Deibel 2006; Roberge 2006; Belloni 2006). Market studies are coming to similar conclusions (Arthur de Little 2006; Pieters et al. 2006). A paradigm change in plant engineering is quoted, governed by time-to-market (too late with products) and standardization issues (Deibel 2006). Here, microprocess engineering will have a role, based more on plant philosophy than on absolute size.

Two main fields of application have been explored: fine chemistry/pharmacy (Hessel et al. 2004a) and fuel processing (Hessel et al. 2005d). A few investigations target manufacture of bulk chemicals or intermediates. In addition, the production of smart and functional materials is the focus, such as pigments (Wille et al. 2004), encapsulated delivery

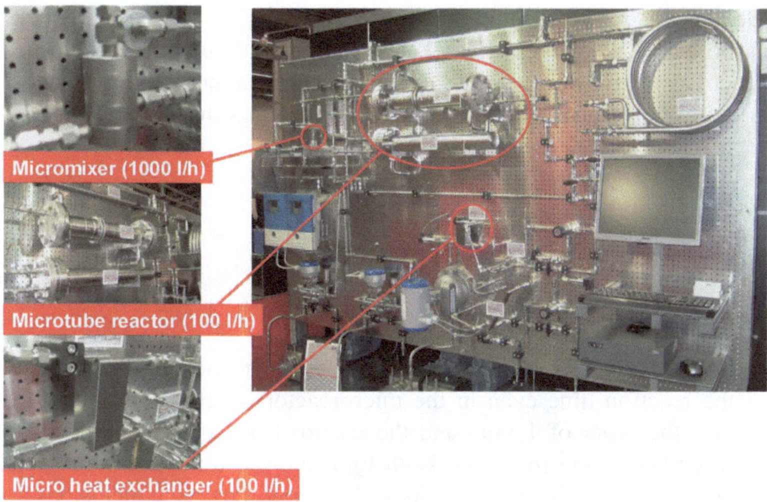

Fig. 1. Demonstration of microprocess technology plant for fine chemistry at IMM

systems (Wischke et al. 2006; Freitas et al. 2003), liquid-crystalline droplets (Fernández-Nieves et al. 2006), or chromatography beads (Kubo et al. 2006), just to name a few. Future investigations will cover personal care, household, and food applications.

In the context of this paper, only the microdevices and the microprocess engineering for uses in fine chemistry and pharmacy are considered. The view will be process-based and deduced from a rough economic calculation of how mature the technology is and what needs to be done to promote it further.

3 Cost Analysis for General Judgment of Microreactor Competitiveness

3.1 Case of Low Plant Costs: Kolbe-Schmitt Synthesis

Fine-chemical processes convert raw materials to higher-value materials with a total annual value much higher than the costs of a typical microprocessing plant and its engineering. To illustrate this, the example of the aqueous Kolbe-Schmitt synthesis from resorcinol to 1,4-dihydroxy carboxylic acid is taken (Hessel et al. 2005b), although the range of respective materials costs in fine chemistry and in microreactor plant cost is quite wide and it cannot be taken for granted that this example is representative.

OH OH $\xrightarrow{KHCO_3\,(aq)}$ COOH OH OH

Nonetheless, this reaction is certainly not at the most profitable edge, as the reaction time even in the microreactor is ultimately not short, still on the order of 1 min, and the reactor load is further limited by dissolubility of the reactants. Both limit the reactor space–time yield and thus decrease the productivity so that the following conclusions do not at all comprise a best-case scenario.

As shown in Fig. 2, the raw materials costs for the Kolbe-Schmitt synthesis amount to approximately 40 €/kg, the target product amounts to about 124 €/kg (U. Krtschil et al., unpublished data). A five-tube microprocessing plant, which is a realistic numbered-up forecast from experimental kinetic data from a single-reactor operation, can achieve about 4.4 t/a, which corresponds to process volumes rated to reactants and products of 176 and 546 k€/kg. The cost of a five-reactor microprocessing plant for the Kolbe-Schmitt synthesis amounts to less than 30 k€. The capital costs total 1.12 €/kg, if a 10-year lifespan and 7-year amortization are assumed. It is evident from these numbers that the costs of a microreactor plant are not prohibitively high, although a complete cost analysis needs to incorporate much more data. This is further corroborated by an analysis of the waste disposable, which amounts to 15 €/kg for the Kolbe-Schmitt synthesis. Since it has been demonstrated many times that microprocess engineering can enhance selectivity significantly, this share of the costs should also be reduced substantially, further improving the costing balance in favor of the microreactors. This could not be demonstrated for the Kolbe-Schmitt synthesis of resorcinol; however, for the same synthesis with phloroglucinol, a 20% increase in yield was found (Hessel et al. 2007).

Base case: high-p,T	µ-REAC PLANT – 1 reactor; 4.4 t/a; 10 l/h 45% selectivity; 200°C / 40 bar (‚high-p,T')	92.10 € / kg
Selectivity	µ-REAC PLANT – 1 reactor; 4.4 t/a; 10 l/h 70% selectivity; 200°C / 40 bar (‚high-p,T')	78.43 € / kg
External numbering-up	µ-REAC PLANT – 10 reactors; 44 t/a; 100 l/h 45% selectivity; 200°C / 40 bar (‚high-p,T')	57.47 € / kg
Further PI by high-p,T	µ-REAC PLANT – 1 reactor; 44 t/a; 100 l/h 45% selectivity; 200°C / 40 bar (‚high-p,T')	56.95 € / kg
Without high-p,T	µ-REAC PLANT – 1 reactor; 0.01 t/a; 0.025 l/h 45% selectivity; 100°C / 1 bar	17,352.52 € / kg
Bench-marking	BATCH-REACTOR PLANT – 1 l; 0.27 t/a 45% selectivity; 100°C / 1 bar (‚Reflux')	985.18 € / kg
Bench-marking	BATCH-REACTOR PLANT – 20 l; 4.3 t/a 45% selectivity; 100°C / 1 bar (‚Reflux')	107.05 € / kg

Fig. 2. Cost analysis for the aqueous Kolbe-Schmitt synthesis for the high-p,T base case

The central elements determining the costs of the microprocessing plant for the Kolbe-Schmitt synthesis are the peripherals such as two pumps, a chilling unit, and a thermostat, while the microreactor equipment includes a micromixer, a capillary as a reactor module, and a micro-heat exchanger for temperature quenching (Hessel et al. 2005b). There are certainly more complex plant process flows for other reactions at a comparable throughput, for instance with many more and more costly microstructured reactors, as well as more and more costly peripherals, while in this example the pumps are the dominating capital cost factor. The sulfonation of aromatics such as toluene is a case in point, which has a gas–liquid reaction with the dedicated and costly falling film microreactor as the central step and requires several downstream microprocessing operations for reaction completion, documented in Müller et al. (2004). The number of pumps here is increased to at least six, dependent on the exact configuration. In addition, control equipment was not considered in the cost analysis for the Kolbe-Schmitt synthesis, which can substantially increase the plant costs. Taking this into account and following the price scheme for microreactors and peripheral equipment as noted for the Kolbe-Schmitt synthesis, microprocessing plants in fine chemistry may cost as much as 500 k€, just to give a figure, although there is no real upper limit here. Nonetheless, this rough estimation should serve here to demonstrate that microprocessing plant costs for fine chemistry themselves have a large spread from low costs to be in the order of the product value per year, in other words, substantial. In the following, first the common case of a micromixer-reactor plant, as given for Kolbe-Schmitt synthesis, is considered, with low plant costs, and then a case needing a more advanced plant for large-scale production is given.

Since the capital costs the Kolbe-Schmitt synthesis are thus competitive with existing technology, the operational costs of such novel continuous synthesis need to be determined. The energy and production site costs were found to be negligible. Therefore, the only uncertain cost share remains the labor costs. This is certainly the most difficult parameter in the cost analysis, since the share of labor time of the operator per microreactor plant cannot be theoretically calculated, but needs to be determined by future experience. At present, an operator share based on industrial experience with batch plants was taken as a first estimation. In this way, labor costs amounted to 36 €/kg, i.e., they were nearly as high as the raw material costs. The main figure for calculating such labor costs, besides the number of plants controlled by one operator, is the productivity of plant operation, which inversely goes along with the reaction time, i.e., which is proportional to the speed of the reaction.

3.2 The Case of High Plant Costs, Hypothetical Reaction

The relevant shares in the cost calculations change when microprocess technology approaches large-scale chemical production (U. Krtschil et al., unpublished data). Reactor fabrication at a competitive cost becomes crucial and is among the main cost drivers. The corresponding costs here go with the overall size of the microstructured reactor and with its number.

A hypothetical reaction with assumed properties was taken, since no real-case reaction was available for demonstration. Liquid-phase operation for a homogeneously catalyzed pseudo-first-order reaction was considered. Calculations were made using the Arrhenius equation with the pre-exponential factor of 4×10^6 l/(mol s) and the activation energy 5×10^4 J/mol. A reaction velocity constant of 0.01 1/s was assumed at a temperature of 100°C. In the next step, the temperatures corresponding to reaction velocity constants of 0.1, 0.05, and 1 1/s were calculated. The reaction times for a 95% conversion were derived in this way, and using these data microstructured plate reactors for use in large-scale chemical production were dimensioned. The overall capacity was set to 10,000 tons of product per year. One microstructured reactor unit has a width, height, and length of 0.25 m × 0.35 m × 0.5 m. Today's microfabrication techniques for steel can still machine such extended areas; however, this is at the upper edge of what can be commercially processed. Still larger reactor units would demand a technology change to mass manufacture techniques such as rolling for microstructuring steel plates (see Fig. 3) and brazing for their interconnection. To comply with other calculations of these larger microstructured reactors, this was taken as the cost basis in this case as well (and not the current practices in chemical etching and laser welding). Channel dimensions were set to 0.5 mm deep, 1 mm wide, and 0.5 m long to meet the capacity requirements. Reaction times ranged from 3 s to 5 min, depending on the temperature.

Figure 4 shows the impact of process intensification for this hypothetical case. With a temperature increase of only 41°C, the number of reactors for such comparatively big units is reduced from 20, hardly feasible in view of costs and process control, to four, feasible for the same reasons. Thus, the costs decrease by almost a factor of five (not exactly, since fixed costs have a small share). Another 21°C temperature increase halves the number of reactors again, and at 249°C, which is 149°C higher than the base temperature, an equivalent of 0.2 microstructured reactors is needed. This means that practically one microstructured reactor is taken and either reduced in plate number or in the overall dimensions. The costs of all microstructured reactors scales largely with their number; only at very low numbers do fixed costs for microfabrication lead to a leveling off of the cost reduction.

In another cost analysis, a commercial fine-chemical synthesis using microprocess technology, developed and undertaken by the AzurChem Company, was investigated (Krtschil et al. 2006). Using real-life data,

Fig. 3. Roll for mass fabrication of microstructured steel plates. The endless sheet can be moved at least at a velocity of 1 m/min, i.e., for a plate length of 10 cm and rolling four microstructured units in parallel, 2,400 plates/h can be manufactured. Structural dimensions of 500 μm have so far been realized; the technique is expected to yield microstructures at a100-μm characteristic width

the findings of the Kolbe-Schmitt synthesis were confirmed with the example of phenyl boronic acid synthesis. Here, selectivity has a much higher impact, as both starting and product materials are very costly. In addition to the real case, three other hypothetical microflow cases and one hypothetical batch case were considered to give an outline of how process protocol variations might have an impact (see Fig. 5). This includes a five- and tenfold capacity by process intensification (novel chemistry concepts) or a tenfold parallel microreactor operation (numbering-up). In addition to showing their impact on the cost structure, the resulting earnings and total costs are given.

Without further detailing such analysis, reported comprehensively elsewhere, some simple rules for cost-efficient microprocess technology fine-chemical synthesis can be given. The first rule is to select fast

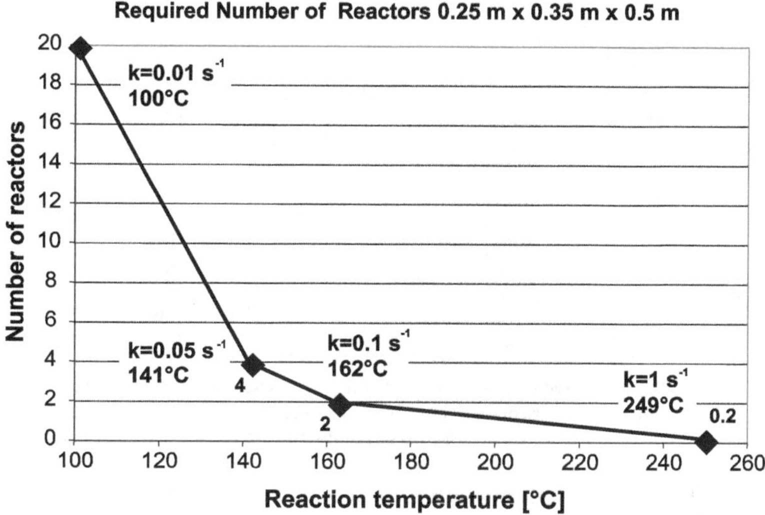

Fig. 4. Impact of process intensification on a hypothetical homogeneously catalyzed liquid reaction. Number of parallel operating reactors depending on the reaction temperature

chemical processes. An investigation by the company Lonza has shown that approximately 44% of all industrial processes fulfilled this criterion within 1 year; 17% when considering the current practical limitations (Roberge et al. 2005). On the other hand, this means that one must question whether there is unreleased potential or whether the rest of the reactions are perhaps not suitable. Thus, the second issue is to check if and how one can transform otherwise slow processes into fast ones. The analysis also showed the need to select high-value reactants and products, which is given for expensive fine chemicals and pharmaceutical products and for most advanced and functional materials. In turn, microfabrication costs are often negligible (but only in case of success). Preliminary cost analysis calculations show, however, that for bulk chemistry, fuel processing, consumer goods, etc., reactor and energy costs have an impact. Mass manufacture of microreactor plates is the key here to cost-efficient reactor provision.

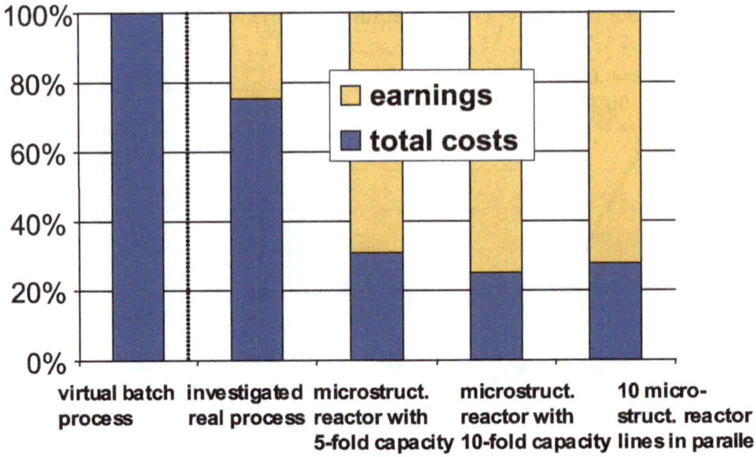

Fig. 5. Earnings and total costs for a commercial microreactor process, for the real-life micro-flow case, three processes intensified or numbered-up micro-flow cases, and a batch benchmark case (Courtesy of Swiss Chemical Society)

4 New Process Windows: Speed up Microreactor Chemistry

The arguments given above underline that using a microreactor without a high degree of parallelization or without speeding up the reaction may not be economically competitive. On the other hand, taking these two measures into account, there is a wide range of microreactor cost performance. Thus, the lesson to be learned from this is that one should not passively use microreactors by applying established chemical protocols, but rather actively re-think the chemical protocols themselves to achieve a significant improvement in technical and cost performance. This seems to contradict the general rule of process intensification and microprocess technology requiring that the reactor be designed around the reaction rather than adapting the reaction to make it compatible with the reactor. However, basically this is extending that rule by saying that in order to tap the full potential of microprocess technology, a revision and re-thinking of the chemistry itself is required, i.e., it needs to be shifted to the more extreme edges of the processing window, for example, in terms of temperature, pressure, concentration, shortness of residence time, etc. Basically, it is not surprising that the chemistry developed over centuries for slowly processing batches is not ideal for extremely fast-acting reactors.

For this reason, a chemical protocol design strategy called initially novel chemistry and re-named recently new process windows (Hessel et al. 2005c) was proposed, including three sources of information, which are prior scattered works in the field of organic chemistry (including some geochemistry papers; Siskin and Katritzky 1991), specific research in the field of the emerging interdisciplinary fields such as microprocess technology or microwave synthesis, and some predictions we have developed to complete the picture. The cornerstones of the new process windows concept are:

- Direct routes from hazardous elements
- Routes at increased concentration or even solvent-free
- Routes at elevated temperature and/or pressure
- Routes mixing the reactants all at once
- Routes using unstable intermediates
- Routes in the explosive or thermal runaway regime
- Process simplification – e.g., routes omitting the need of catalysts or (complex) separation

In particular, inspired by our own work was the second concept, termed high-p,T, which stands for combining high temperatures (to accelerate chemical processes) and high pressures (to prevent the solution in the microreactor from boiling and keep it in the liquid phase). Although this is a trivial measure and certainly was done before using conventional technology, microprocess technology adds a low technical expenditure and improved safety when exploring such explosive or hazardous regions in the process window. Thus, while being possible in the past, microreactors largely expand the applicability of high-p,T processing. The same is utilized in microwave processing, showing an increasing number of continuous applications, and has been named encased processing.

In this way, typical organic protocols can be shifted from about 130°C maximal operational temperature (depending mostly on the choice of solvent, with the exception of the less frequently used DMF and DMSO with lower boiling points) to 200°C and higher. Typical pressures applied are 50 bar and higher to ensure single-phase operation with a large safety margin.

{50-130°C; 1 bar; 2-8 h}

↓

{>200°C; >50 bar; <1 s}

'alike gas-phase chemistry'

An idea on the impact of high-temperature processing was provided by calculating the time needed for a 90%-conversion degree based on the Arrhenius equation, assuming a first-order reaction with the enhancement factor $A = 4 \times 10^{10}$ mol^{-1} s^{-1} and the activation energy E_a=100 kJ mol^{-1} (Kappe and Stadler 2005). This shows that order of magnitude changes can be achieved with practicable increases in reaction temperature. With simply an increase of roughly 100°C, reactions formerly taking days are shortened to hours and those formerly taking hours need only a few seconds. This was verified by microwave processing on its own and by combined use of microwaves and microflow processing. It stands to reason that such considerations only target conversion (and reaction rate consequently) and do not take into account selectivity which may dramatically decrease at high temperatures by decomposition or side reactions. To control the latter, shortening residence times is one approach.

The high-p,T effect was demonstrated at the example of the aqueous Kolbe-Schmitt synthesis of resorcinol (see formula given above) by performing microflow processing on its own, using a pressurized capillary, if needed equipped with a micromixer upfront and a micro-heat exchanger for temperature quenching downstream (Hessel et al. 2005b).

Table 1

Temperature [°C]	27	77	127	177	227
Rate constant k [s^{-1}]	1.6×10^{-7}	4.8×10^{-5}	3.5×10^{-3}	9.9×10^{-2}	1.43
Time (90% conversion)	68 days	13.4 h	11.4 min	23.4 s	1.61 s

Experiments were conducted at a pressure of 40–70 bar, a temperature of 100–220°C, and reaction times of 4–390 s. It was cross-checked and confirmed that the textbook recommended reaction time of 2 h at least roughly constitutes the actual kinetically needed reaction time at the standard batch conditions of 100°C and 1 bar (which is usually not the case; over-reaction is common in organic synthesis).

Using the high-p,T microreactor processing, the Kolbe-Schmitt synthesis was completed within less than 1 min at comparable yields, i.e., a reaction time reduced by a factor of approximately 2,000 was achieved (see Fig. 6). This corresponds to an increase in space-time yield by a factor of 440.

As can be expected, the high-temperature processing runs the risk of enhancing side and consecutive reactions. Decarboxylation of the main product was found and increases with temperature (see Fig. 7). This is illustrated at the example of the synthesis of 2,4,6-trihydroxy benzoic acid from phloroglucinol, as this molecule is even more sensitive to thermal destruction due to the enhanced electron richness of the aromatic core by presence of a third hydroxyl group (Hessel et al. 2007).

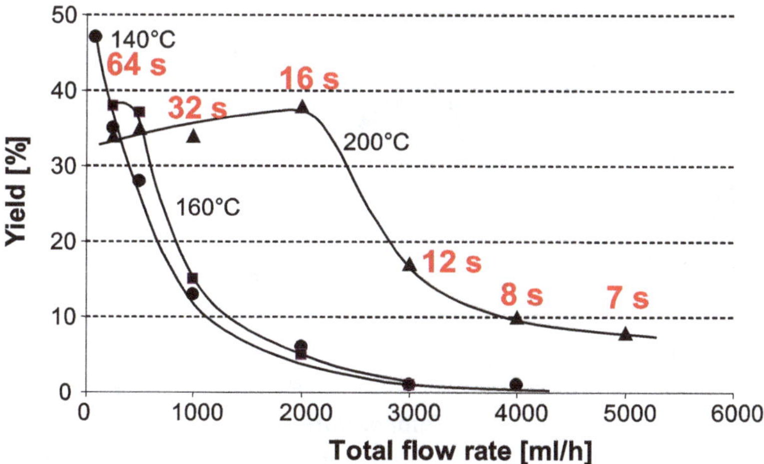

Fig. 6. Process intensification of the aqueous Kolbe-Schmitt synthesis of resorcinol using high-p,T processing. (Courtesy of ACS)

Thus, there is a small process window where high-p,T processing is effective and process control may have an important role here to maintain processing at optimal conditions.

Figure 8 illustrates how delicate the impact of residence time and temperatures is. Optimal performance is only achieved at short times not exceeding a few minutes and overshooting the temperature by roughly 10°C can already notably decrease the yield.

Similar process intensification was obtained for the Michael addition with α, β unsaturated amines and esters (Löwe et al. 2006).

Microstructured reactors made it possible to take advantage of the fast kinetics of the Michael addition reaction, while still ensuring efficient heat removal and avoiding thermal overshooting. Yields ranging from 95% to 100% were obtained, at reaction times of 1–10 min instead of the 24-h batch operation; a high-p,T operation was established. High

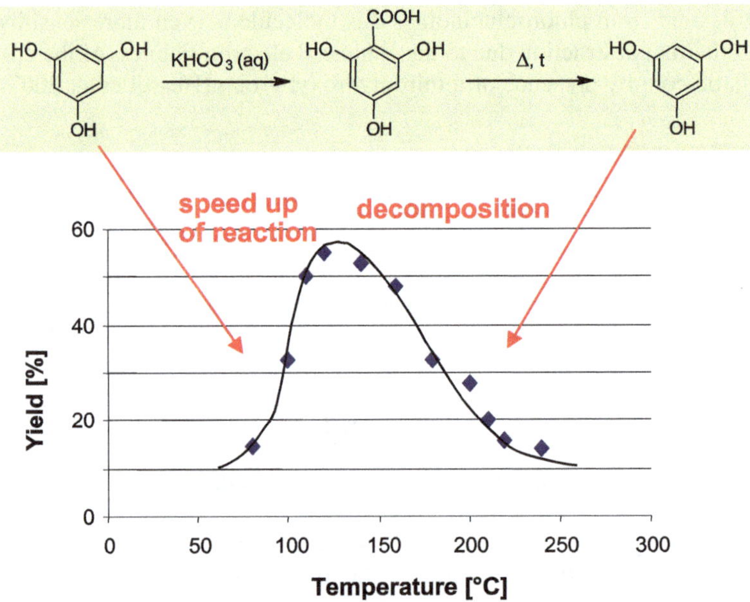

Fig. 7. Process intensification and adverse thermal destruction effects of the aqueous Kolbe-Schmitt synthesis of phloroglucinol using high-p,T processing. (Courtesy of Wiley-VCH)

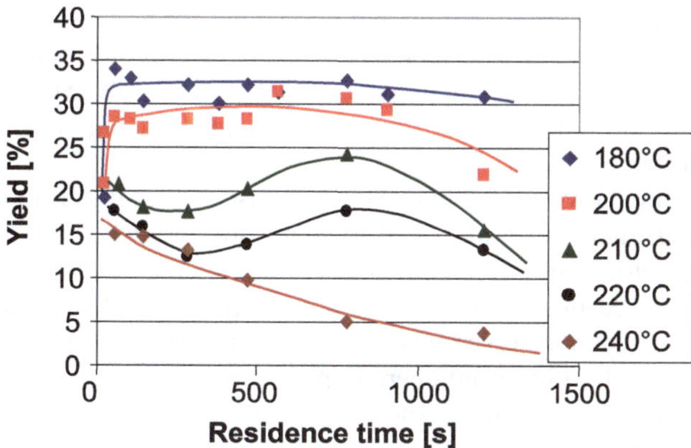

Fig. 8. Adverse thermal destruction effects of the aqueous Kolbe-Schmitt synthesis of phloroglucinol using high-p,T processing. (Courtesy of Wiley-VCH)

$$RR'NH + H_2C{=}CH{-}R'' \rightleftharpoons RR'N{-}CH_2{-}CH_2{-}R''$$

with R, R' = alkyl-
R'' = -COOR, -CN

productivities were achieved, for example, 100 g piperidino-propionitrile in 13 min ≈ 9 l/day.

5 Some Thoughts on Scale-out of Microprocess Technology

Different methods for scale-out have been collected and presented (Löb et al. 2006a). The two main methods are a smart scale-up, at best after having intensified the process, or the internal numbering-up concept. The first route was shown long ago by the Merck production case for an organometallic reaction, changing from micro- to mini-mixers (Krummradt et al. 2000). This route seems to be feasible, as long as there is single-phase operation and a potential for process intensification. Then, by use of one microstructured device, most often a mixer, connected to a tube, production can be achieved simply. Since micromixers are short, the pressure drop is low and, in turn, high throughputs can be achieved.

There are, however, known laboratory case studies in fine chemistry that will demand much more dedicated reactor design when moving to a pilot reactor design. This holds true, for example, for reactions that need advanced thermal management, multiphase reactions, and relatively slow reactions (even when intensified). The corresponding pilot reactors will be highly numbered-up, highly integrated, and/or have a long flow axis. This means the corresponding pilot reactors will have comparatively large dimensions overall and may become costly due to a comparatively huge mass of construction material and large-area microfabrication. At this point of development, these pilot reactors are just beginning construction.

In the following, the reasons for the higher architectural demands are exemplified at two laboratory examples for liquid–liquid Suzuki coupling and imidazol-based ionic liquid synthesis and can only partly be shown in the example of liquid–liquid and gas–liquid processing scale-out.

5.1 Liquid–Liquid Dispersion for Suzuki Coupling: A Redispersion Reactor

The dispersion of organic solvents and aqueous solutions, commonly found in phase-transfer reactions, in a continuous mode by the simple use of microstructure mixers is a difficult operation, since in most cases fast coalescence of the created droplets may be found, possibly slowing down the reaction. One solution is the repeated break-up of droplets. A redispersion mixer-reactor with four dispersion stages was developed. The basic function was proven with dyed aqueous solution with the surfactant sodium dodecyl sulphate and heptane, as shown in Fig. 9. The coarse droplet system, premixed using an external micromixer (not shown), is dispersed more finely from stage to stage (see Fig. 9). The stage-wise redispersion leads to an extended reactor length, and accordingly the flow per channel or unit is kept relatively low, leading to pilot

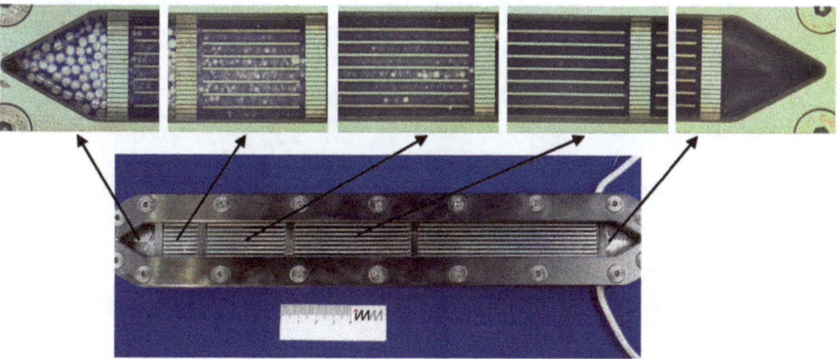

Fig. 9. Stage-wise breakup of a coarse droplet mixture in a redispersion microreactor. (Courtesy of Wiley-VCH)

designs with a higher degree of numbering-up and with extended reactor length.

The redispersion microreactor is applied for the liquid–liquid polycondensation to yield an OLED material by multiple Suzuki coupling. As the initial test reaction, the following single Suzuki coupling is currently being explored in the liquid–liquid system made from water/dioxane/toluene.

$$C_6H_5\text{–}B(OH)_2 + Br\text{–}C_6H_5 \xrightarrow{Pd(P(o\text{-tolyl})_3)_4,\ K_3PO_4} C_6H_5\text{–}C_6H_5$$

Measurements with the latter real-case solvent system give a more complex picture. Even for the one-plate design, achieving uniform flow conditions in each channel is an open issue. Preliminary experiments show that high-temperature, high-pressure operation gives better flow uniformity, which is also advantageous to speed up the polycondensation reaction (see Fig. 10).

5.2 Solvent-Free Ionic Liquid Synthesis with Large Heat Release: Capillary or Integrated Reactors

An imidazol-based ionic liquid synthesis was carried out under solvent-free conditions, simply bringing the two liquid reactants into contact

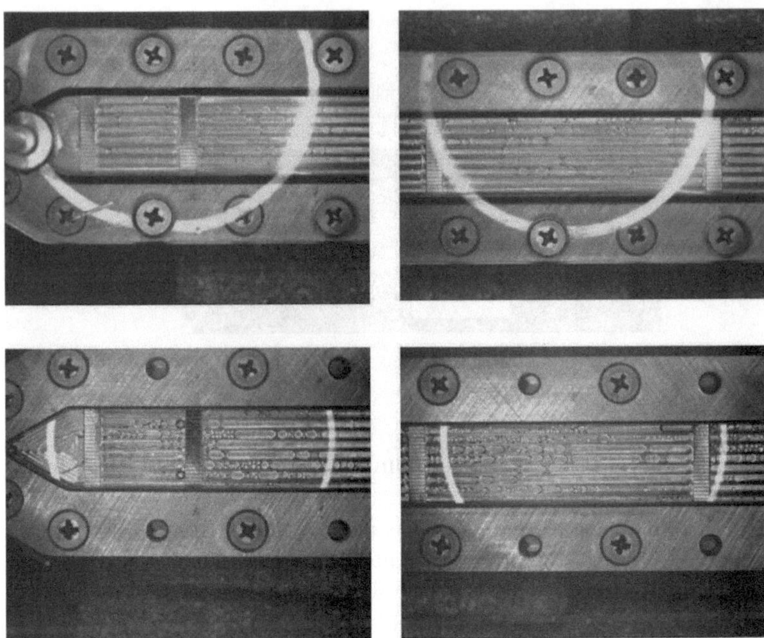

Fig. 10. *Top images*: Improved flow distribution of the liquid–liquid slug flow at 80°C and 5 bar at the reactor inlet (*left*) and the reactor center (*right*). Less regular flow distribution of the liquid–liquid slug flow at 20°C and 1 bar at the reactor inlet (*left*) and the reactor centre (*right*)

(Löb et al. 2006b). Because of the high reactor load and large exothermicity of the reaction, the heat releases are quite strong. In-line temperature measurements in a micromixer-capillary set-up cooled by a temperature bath confirm the existence of hot spots of much more than 50°C in the capillary (see Fig. 11).

This can be overcome by using a multiscaled capillary design or by integrating heat exchange and reaction channels within one microreactor unit, both at the expense of an increasing drop in pressure. Thus, pilot-sized reactors will be highly numbered up or highly integrated. This additional expenditure in reactor design is, however, outweighed

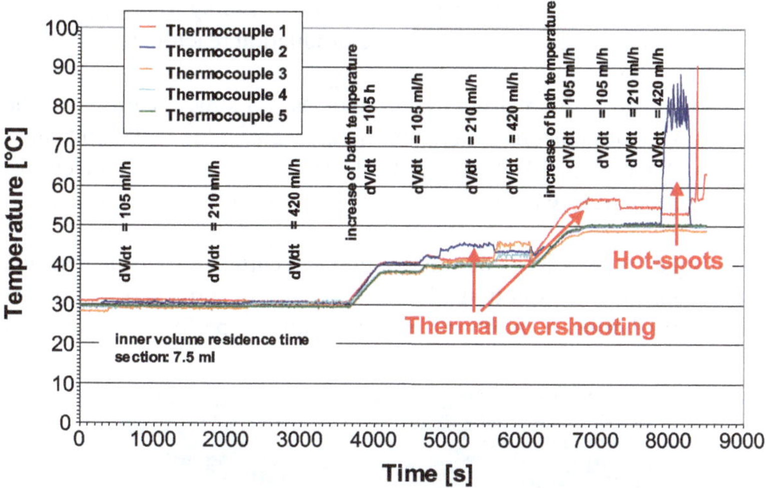

Fig. 11. Real-time, in-line temperature monitoring for an ionic-liquid synthesis when varying external bath temperature and volume flow

by the process intensification through fast, high-temperature, solvent-free continuous processing.

Yoshida et al. developed similar numbered-up capillary pilot reactors for free radical polymerization (Iwasaki et al. 2006). The capillaries were either arranged in parallel fashion similar to conventional multitube reactors or consecutively branched by multiport valves.

5.3 Gas–Liquid Processing Scale-out: Falling Film Microreactor

The falling film microreactor has been applied to a number of different reactions and has been characterized for mass transfer at the laboratory scale. In a running German BMBF project with Degussa as the industrial partner, scaled-out falling film microreactors are being developed; they have a tenfold larger structured area (150 cm^2) and a correspondingly larger film surface compared to their laboratory counterparts (Vankayala et al., unpublished data). One pilot microstructured reactor refers to the numbering-up concept with a more compact arrangement of an increased number of microchannels on a cylindrical body (see

Fig. 12). External numbering-up by flow manifolding to separate reactor units is here one concept for future production reactors with another tenfold increase in structured area.

Another pilot microstructured reactor uses a smart scale-up of the reactor plate dimensions by a factor of 3.3, each for the width and channel length (see Fig. 13). Stacking further plates is here one concept for future production reactors with another tenfold increase in structured area.

Using an industrial oxidation reaction and carbon dioxide absorption in aqueous alkaline solutions, it was validated in some sets of process conditions that such pilot reactors show similar performance as their

Fig. 12. Cylindrical falling film microreactor for pilot operation at tenfold capacity increase as compared to the laboratory falling film microreactor

laboratory counterparts. At best, a selectivity of 90% at a conversion of up to 80% was achieved. By operating at 10 bar, increases in yield by 5%–15% (within the flow rate range investigated) were found in 1,200-μm-wide microchannels, as compared to atmospheric pressure, which again demonstrates the need for intensified processing when using microprocess technology.

Future issues for an improved pilot reactor design remain achieving and confirming the same falling film profile in the microchannels for the laboratory and pilot reactor. The film shapes are complex and theoretical predictions seem to fail to accurately describe them. It stands to reason that such thin films have a dedicated dependence on the details

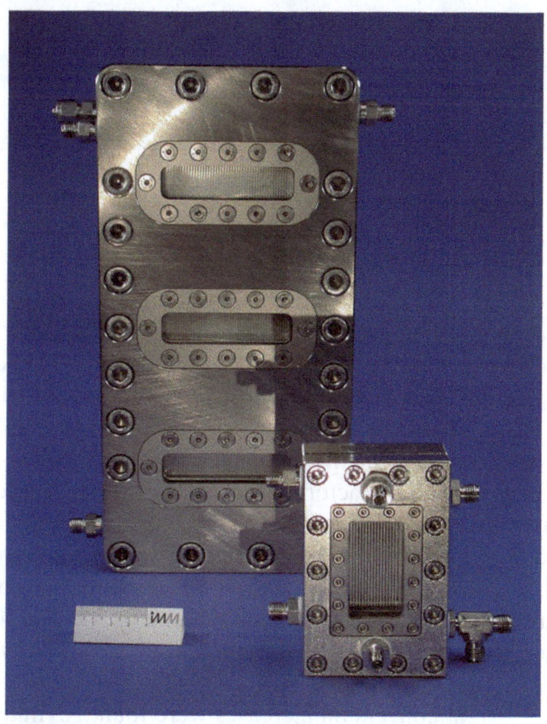

a

Fig. 13. a,b Scaled-out large falling film microreactor for pilot operation at tenfold capacity increase compared to the laboratory falling film microreactor

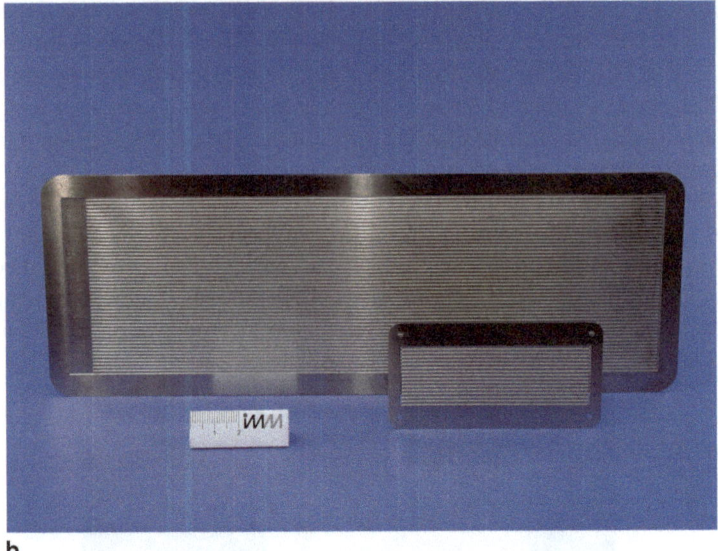

b

Fig. 13. a,b (continued)

of the microchannel encompassing them. In their current version, laboratory and pilot falling film microreactors may differ in the surface energies and cross-sectional shapes of their microchannels because of the need to apply different microfabrication technologies or to have varying qualities of microfabrication for the same technology. While chemical etching is used to make the microchannels at the laboratory-scale plates of the falling film microreactor, microelectro-discharge machining was the adequate preliminary approach to the cylindrical pilot design. For the scaled up pilot design, chemical etching was suitable to make both the small laboratory and the large pilot plates. Still, as etching is done by a spray technique from a punctual source to wet and abrase a flat body, there may be differences in microstructured quality at varying plate scales.

Actually, slight conversion increases were found: cylindrical FFMR B large-plate FFMR B standard FFMR. This is due to different settings of residence time and film thickness, which come from the above-mentioned slightly different geometric specifications. Film thickness is the central, often unknown parameter for scale-out, as standard correlations such as the Kapitza, Feind, or Nusselt equation are only accurate by one order of magnitude and the demand for additional experimental determination is difficult in routine operation. Such effects of microstructuring on channel geometry and surface energy are under investigation.

6 Industrial Production Cases

6.1 Nitroglycerine Production Plant

A nitroglycerine microprocessing plant, developed by IMM for the Chinese class-A company Xi'an, is an example of the smart scaled-out concept, showing that just one caterpillar micromixer is sufficient for the continuous production of pharmaceutical-degree nitroglycerine, with a throughput of approximately 15 kg/h (Thayer 2006).

$$\text{HOCH}_2\text{CH(OH)CH}_2\text{OH} \xrightarrow{HNO_3\,/\,H_2SO_4} \text{O}_2\text{N-O-CH}_2\text{CH(O-NO}_2\text{)CH}_2\text{-O-NO}_2$$

The manufactured nitroglycerine will be used as a drug for acute cardiac infarction. Therefore, the product must be made under GMP conditions and the quality of the product has to meet the highest standards. The microstructured reactor plant consists essentially of three main parts (see Fig. 14): mixing sulfuric acid with nitric acid, with highly concentrated fuming liquids, the reactor and the phase separation, and washing and purification devices. On demand, glycerine and the acid mixture are fed separately into a microstructured reactor where mixing occurs within milliseconds. The high surface-to-volume ratio of these new types of reactors ensures an immediate transfer of heat released by the reaction. Increased safety compared to conventional processing is achieved by the small volume hold-up in the microstructured reactor. In summary, higher yields, better product quality, increased safety, and a reduction in environmental hazards are the main advan-

Fig. 14. Nitroglycerine production microprocess plant with staff at industrial site in Xi'an, China

tages of applying microstructured reactors to this type of chemical reaction.

6.2 Radical Polymerization Plant

A pilot plant for free-radical polymerization was developed at Kyoto University (see Fig. 15) and transferred to the industrial site at the Idemitsu Kosan Company (Iwasaki et al. 2006).

Two microprocess technology concepts were evaluated based on different multicapillary designs (see Fig. 16).

For the superior performance design with consecutive, multiply furcated capillaries using conventional manifolds, flow distribution quality was determined. Polydispersity indices for eight capillaries ranged from 1.66 to 1.71, i.e., it showed a low spread, demonstrating that flow distribution was reasonably achieved. The molecular number average was

Fig. 15. Radical polymerization reaction pilot microprocess plant (Courtesy of ACS)

18,000 at a low yield of approximately 6%. Stable process operation over a period of 7 days was demonstrated by showing an almost constant degree of polymerization and polydispersity index.

6.3 Grignard Exchange Reaction Plant

Another pilot plant, developed by at Kyoto University, was used for the Grignard exchange reaction at the same productivity as the batch reactor (10 m^3) by adding only four microflow systems of the present scale (Wakami and Yoshida 2006). A multitubular reactor is the core element of the microprocessing plant (see Fig. 17). Stable yields of approximately 95% could be demonstrated for a 24-h run.

6.4 Swern-Moffat Oxidation Plant

Figure 18 shows a pilot plant with a capacity of 10 t/a built for Swern-Moffat oxidation (Kawaguchi et al. 2006). The microreactor includes a series of rapid mixing functions for mass production. The yield of

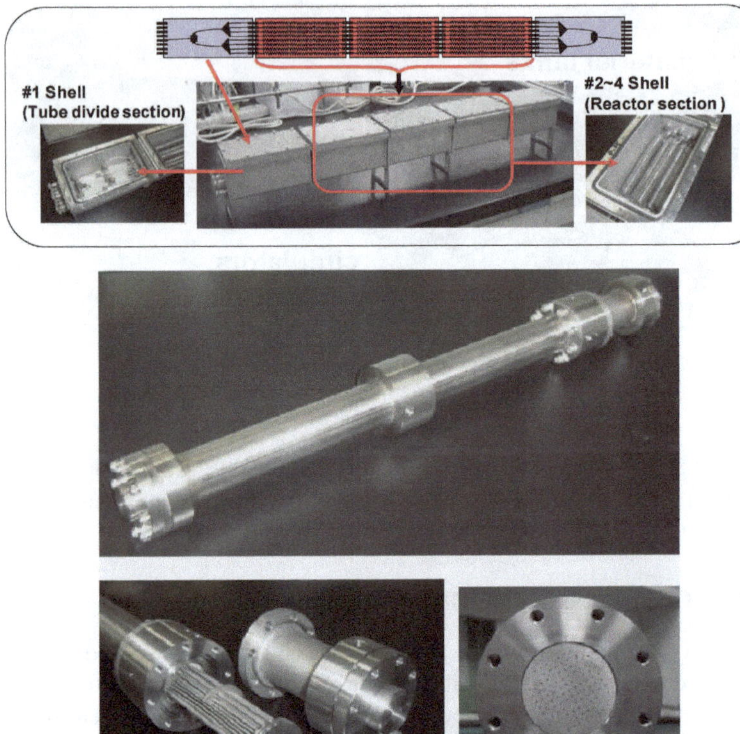

Fig. 16. a,b Two parallel microcapillary connection concepts: tube-and-shell multitubular and branched tube arrangements (Courtesy of ACS)

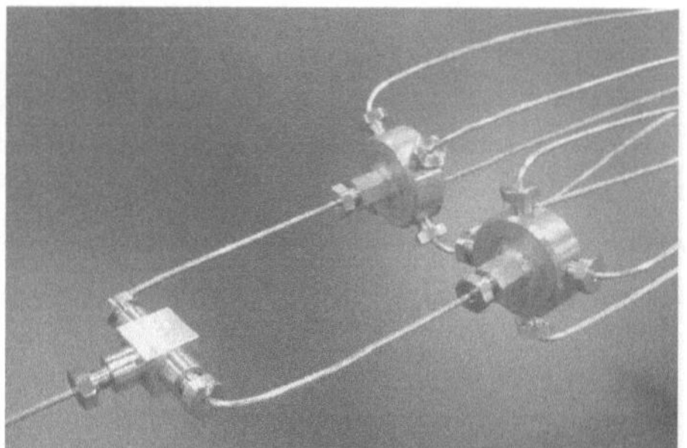

b Tube branch structure using simple couplers

Fig. 16. a,b (continued)

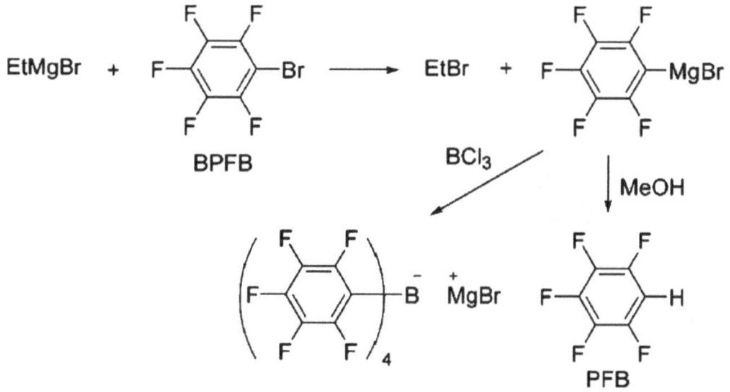

EtMgBr + BPFB (F-substituted pentafluorobromobenzene: C_6F_5Br) → EtBr + C_6F_5MgBr

C_6F_5MgBr + BCl_3 → $(C_6F_5)_4B^-\ {}^+MgBr$

C_6F_5MgBr + MeOH → C_6F_5H (PFB)

the batch process is 83% at –70°C, whereas the microchemical process achieved a yield of 88% at 20°C. The pilot plant could be operated stably for a long run with the same product yield as the laboratory experiment.

6.5 Yellow Nano-Pigment Plant

Figure 19a shows a pilot plant for the production of a yellow nano pigment with a capacity of 70 t/a, developed and operated by Kyoto University and Fuji Company (Maeta et al. 2006). Figure 19b shows some selected particle-size spectra obtained under various flow conditions.

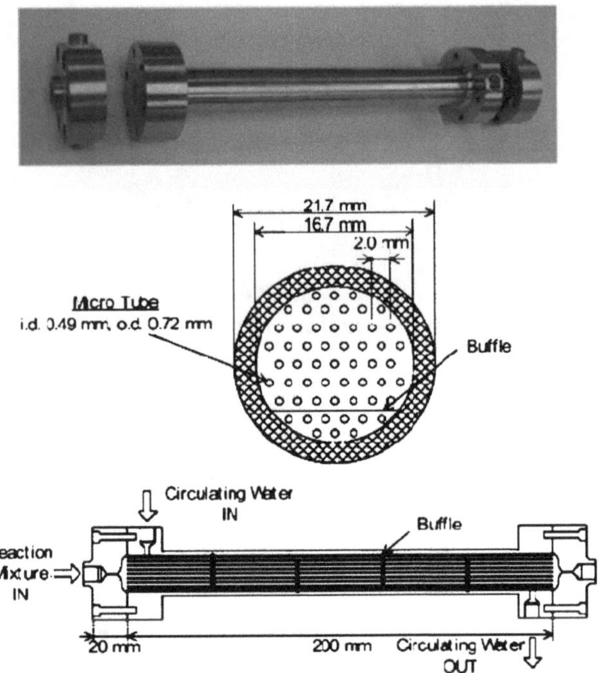

Fig. 17. Multitubular reactor used for the Grignard exchange reaction pilot microprocess plant. (Courtesy of ACS)

6.6 Polymer Intermediate Plant

A customized microstructured reactor, approximately shoebox-size, for manufacturing a high-value polymer intermediate was developed at Forschungszentrum Karlsruhe and installed in an industrial environment (FZK press release 13, 2005). During a 10-week production campaign, over 300 tons of a polymer intermediate product were manufactured for the plastics industry. A microstructured reactor, made from a special nickel alloy, 65 cm long and weighing 290 kg, was the central element of this new production plant. With a throughput of 1,700 kg of liquid chemicals per hour, this reactor was able to meet the demands for a chemical production line. It clearly fits the remarks concerning

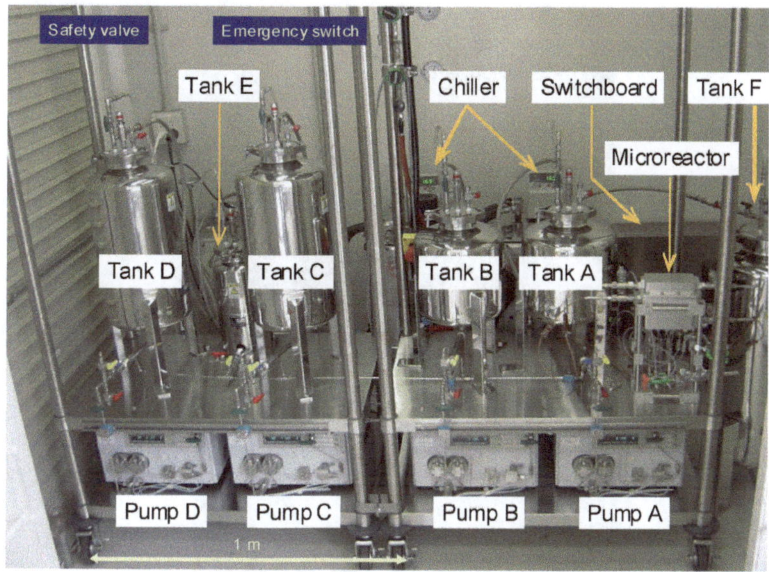

Fig. 18. Swern-Moffats reaction pilot microprocess plant. (Courtesy of ACS)

microstructured reactors made above. The interior of the reactor consist of a multitude of microchannels, while the entire reactor is much bigger and heavier than the term "micro" implies. In this way, a central reaction route in industry has been replaced, which was conducted before in a very large reactor tank.

7 Conclusions and Outlook

Microreactor and microprocess technology has, in some fine-chemical cases, approached commercial applications and become competitive with existing technology. Two main developments are awaited. Firstly, optimizing the process protocol conditions such that the chemistry is set to the limit of the reactor's capabilities in terms of mass and heat transfer. This so-called novel chemistry approach achieves the highest process intensification and can improve the costing of microprocess

a

Fig. 19. **a** Yellow nano pigment pilot microprocess plant

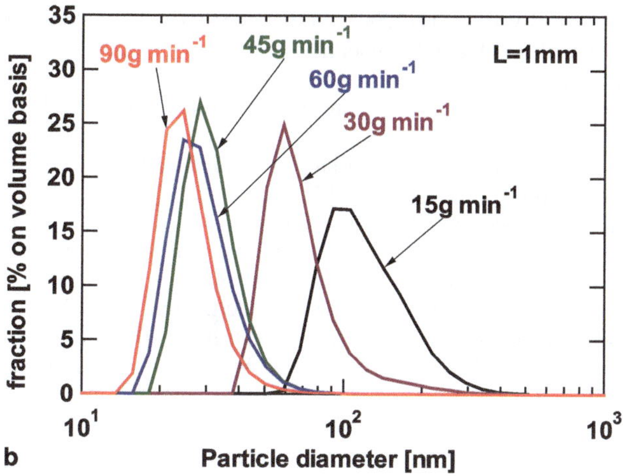

Fig. 19. **b** Particle-size spectra at various volume flows

plants by another order of magnitude. Secondly, new large microstructured reactors need to be built for large-scale applications in general and for more complex operations in particular, such as combined mixing and heat exchange at high mixing sensitivity and/or large heat releases. Here, a paradigm shift in microreactor fabrication using mass manufacture is needed and costing analyses have to be made on a much more complex level. To achieve competitiveness, process intensification is even more necessary than for fine-chemical synthesis. In addition, the costing, the ease, speed, and predictability of scale-out of microprocess technology will largely govern the industrial implementation.

References

Arthur de Little, SenterNovem (2006) Full study presented to platform for chain efficiency (PKE). Building a business case on process intensification

Belloni A (2006) Process 13:64–65

Burns JR, Ramshaw C (2001) The intensification of rapid reactions in multiphase systems using slug flow in capillaries. Lab Chip 1:10–15

Deibel S-R (2006) Anlagenbau: Schneller planen, rascher produzieren. CHE Manager 2:12

Ehrfeld W, Hessel V, Löwe H (2000) Microreactors: new technology for modern chemistry. Wiley-VCH, Weinheim

Fernández-Nieves A, Christobal G, Garcés-Chávez V, Spalding GC, Dholakia K, Weitz DA (2006) Optically anisotropic colloids of controllable shape. Adv Mat 17:680–684

Fletcher PDI, Haswell SJ, Pombo-Villar E, Warrington BH, Watts P, Wong SYF, Zhang X (2002) Tetrahedron 58:4735

Freitas S, Walz A, Merkle HP, Gander B (2003) Solvent extraction employing a static micromixer: a simple, robust and versatile technology for the microencapsulation of proteins. J Microencapsulation 20:67–85

FZK Press release 13 (2005) http://www.fzk.de/fzk/idcplg?IdcService=FZK&node=2374&document=ID 050927. Cited 6 July 2005

Gavriilidis A, Angeli P, Cao E, Yeong KK, Wan YSS (2002) Trans IChemE 80/A:3

Haswell ST, Watts P (2003) Green Chem 5:240

Hessel V, Hardt S, Löwe H (2004a) Chemical micro process engineering—fundamentals, modelling and reactions. Wiley-VCH, Weinheim

Hessel V, Löwe H, Schönfeld F (2004b) Micromixers—a review on passive and active mixing principles. Chem Eng Sci 60:2479–2501

Hessel V, Angeli P, Gavriilidis A, Löwe H (2005a) Gas-liquid and gas-liquid-solid microstructured reactors—contacting principles and applications. Ind Eng Chem Res 44:9750–9769

Hessel V, Hofmann C, Löb P, Löhndorf J, Löwe H, Ziogas A (2005b) Aqueous Kolbe-Schmitt synthesis using resorcinol in a micro-reactor laboratory rig under high-p,T conditions. Org Proc Res Dev 9:479–489

Hessel V, Löb P, Löwe H (2005c) Development of microstructured reactors to enable organic synthesis rather than subduing chemistry. Curr Org Chem 9:765–787

Hessel V, Löwe H, Müller A, Kolb G (2005d) Chemical micro process engineering—processing and plants. Wiley-VCH, Weinheim

Hessel V, Hofmann C, Löb P, Löwe H, Parals M (2007) Micro-reactor processing for the aqueous Kolbe-Schmitt synthesis of hydroquinone and phloroglucinol. Chem Eng Technol 31, (in press)

Iwasaki T, Kawano N, Yoshida Y-I (2006) Radical polymerization using micro flow system. Numbering-up of microreactors and continuous operation. Org Proc Res Dev 10:1126–1131

Jähnisch K, Hessel V, Löwe H, Baerns M (2004) Chemistry in microstructured reactors. Ang Chem Int Ed 43:406–446

Jensen KF (1999) Microchemical systems: status, challenges, and opportunities. AIChE J 45:2051–2054

Jensen KF (2001) Microreaction engineering – is small better? Chem Eng Sci 56:293–303

Joanicot M, Ajdari A (2005) Droplet control for microfluidics. Science 309:887–888

Kappe CO, Stadler A (2005) Microwaves in organic and medicinal chemistry, Vol. 25. In: Mannhold R, Kubinyi H, Folkers G (eds) Methods and principles in medicinal chemistry. Wiley, Weinheim, pp 94–95

Kawaguchi T, Miyata H, Ataka K, Mae K, Yoshida J-I (2006) Room temperature Swern oxidations by using a microscale flow system. Angew Chem Int Ed Engl 44:2413–2416

Kiwi-Minsker L, Renken A (2005) Microstructured reactors for catalytic reactions. Catalysis Today 110:2–14

Kolb G, Hessel V (2004) Micro-structured reactors for gas phase reactions: a review. Chem Eng J 98:1–38

Kubo A, Shinmori H, Takeuchi T (2006). Atrazine-imprinted microspheres prepared using a microfluidic device. Chem Lett 35:588–589

Krtschil U, Hessel V, Kralisch D, Kreisel G, Küpper M, Schenk R (2006) Cost analysis of a commercial manufacturing process of a fine chemical using micro process engineering. Chimia 60:611–617

Krummradt H, Kopp U, Stoldt J (2000) Experiences with the use of microreactors in organic synthesis. In: Proceedings of Microreaction Technology: 3rd International Conference on Microreaction Technology. Springer, Berlin Heidelberg New York, pp 181–186

Link DR, Anna SL, Weitz DA, Stone, HA (2004) Geometrically mediated breakup of drops in microfluidic devices. Phys Rev Lett 92:054503-1-4

Löb P, Hessel V, Krtschil U, Löwe H (2006a) Continuous micro-reactor rigs with capillary sections in organic synthesis—generic process flow sheets, practical experience, and "novel chemistry". Chimica Oggi Chem Today 24:46–50

Löb P, Löwe H, Hessel V, Hubbard SM, Menges G, Balon-Burger M (2006b) Determination of temperature profile within continuous micromixer-tube reactor used for the exothermic addition of dimethyl amine to acrylonitrile and an exothermic ionic liquid synthesis. In: Proceedings of AIChE Spring National Meeting, Orlando, 23–27 April, 2006

Löwe H, Hessel V, Hubbard S, Löb P (2006) Addition of secondary amines to α,β-unsaturated carbonyl compounds and nitriles by using microstructured reactors. Org Proc Res Dev 10:1144–1152

Maeta et al (2006) Proceedings of the AIChE Spring National Meeting. Orlando, 23–27 April, 2006

Markowz G, Schirrmeister S, Albrecht J, Becker F, Schütte R, Caspary KJ, Klemm E (2005) Microstructured reactors for heterogeneously catalysed gas-phase reactions on an industrial scale. Chem Eng Technol 28:459–464

Müller A, Cominos V, Hessel V, Horn B, Schürer J, Ziogas A, Jähnisch K, Hillmann V, Großer V, Jamc KA, Bazzanella A, Rinke G, Kraut M (2005) Fluidic bus system for chemical process engineering in the laboratory and for small-scale production. Chem Eng J 107:205–214

Pennemann H, Watts P, Haswell S, Hessel V, Löwe H (2004) Org Proc Res Dev 8:422

Pieters B, Andrieux G, Eloy J-C (2006) Technologies and market trends in microreaction technology, Chimica Oggi Chem Today 24:41–42

Roberge DM, Ducry L, Bieler N, Cretton P, Zimmermann B (2004) Microreactor technology: a revolution for the fine chemical and pharmaceutical industries? Chem Eng Technol 28:318–323

Roberge DM, Hogan J (2006) A little goes a long way. Nature 442:351–352

Re-dispersion Microreactor for the formation of Liquid-Liquid Dispersions for Use in Polycondensation

Siskin M, Katritzky AR (1991) Reactivity of organic compounds in hot water: geochemical and technological implications. Science 254:231–237

Schubert K, Brandner J, Fichtner M, Linder G, Schygulla U, Wenka A (2001) Microstructure devices for applications in thermal and chemical process engineering. Microscale Thermophys Eng 5:17–39

Thayer AM (2005) Harnessing microreactions. Chem Eng News Coverstory 83:43–52.

Utada AS, Lorenceau E, Link DR, Kaplan PD, Stone HA, Weitz DA (2005) Monodisperse double emulsions generated from a microcapillary device. Science 308:537–541

Wakami, Yohsid J-I (2006) Grignard exchange reaction using micro flow system. From bench to pilot plant. Org Proc Res Dev 9:787–791

Wille C, Gabski H-P, Haller T, Kim H, Unverdorben L, Winter R (2004) Synthesis of pigments in a three-stage microreactor pilot plant—an experimental technical report. Chem Eng J 101:179–185

Wischke C, Lorenzen D, Zimmermann J, Borchert H-H (2006) Preparation of protein loaded poly(D,L-lactide-co-glycolide) microparticles for the antigen delivery to dentritic cells using a static micromixer. Eur J Pharma Biopharma 62:247–253

Xu S, Nie Z, Seo M, Lewis P, Kumacheva E, Stone HA, Garstecki P, Douglas B, Whitesides GM (2005) Generation of monodisperse particles by using microfluidics: control over size, shape, and composition. Angew Chem Int Ed 44:724–728

Ernst Schering Foundation Symposium Proceedings

Editors: Günter Stock
Monika Lessl

Vol. 1 (2006/1): Tissue-Specific Estrogen Action
Editors: K.S. Korach, T. Wintermantel

Vol. 2 (2006/2): GPCRs: From Deorphanization to Lead Structure Identification
Editors: H.R. Bourne, R. Horuk, J. Kuhnke, H. Michel

Vol. 3 (2007/3): New Avenues to Efficient Chemical Synthesis
Editors: P.H. Seeberger, T. Blume

This series will be available on request from
Ernst Schering Research Foundation, 13342 Berlin, Germany